OXFORD MEDICAL PUBLICATIONS

Forbidden Drugs

Forbidden drugs

Understanding Drugs and Why People Take Them

by

PHILIP ROBSON

Consultant Psychiatrist,
Senior Clinical Lecturer,
University of Oxford

Cartoons by
Patricia Broughton

OXFORD NEW YORK TOKYO
OXFORD UNIVERSITY PRESS
1994

Oxford University Press, Walton Street, Oxford OX2 6DP
Oxford New York
Athens Auckland Bangkok Bombay
Calcutta Cape Town Dar es Salamm Delhi
Florence Hong Kong Istanbul Karachi
Kuala Lumpur Madras Madrid Melbourne
Mexico City Nairobi Paris Singapore
Taipei Tokyo Toronto
and associated companies in
Berlin Ibadan

Oxford is a trade mark of Oxford University Press

Published in the United States
by Oxford University Press Inc., New York

A catalogue record for this book is available from the British Library

Library of Congress Cataloging in Publication Data

ISBN 0–19–262430–X (Hbk)
ISBN 0–19–262429–6 (Pbk)

Typeset by Footnote Graphics, Warminster, Wilts
Printed in Great Britain by Biddles Ltd,
Guildford & King's Lynn

For Renée, Harry, and Sophie

That humanity at large will ever be able to dispense with Artificial Paradises seems very unlikely.

Aldous Huxley

The best of life is but intoxication.

Lord Byron

Every form of addiction is bad, no matter whether the narcotic be alcohol or morphine or idealism.

C. G. Jung

I was teetotal until prohibition.

Groucho Marx

Preface

I try my best never to talk about drugs at social gatherings. This is because I have learnt from hard experience that if I do forget my golden rule and get drawn on the subject and my wife is not on hand to turn the conversation to safer matters, the discussion invariably becomes heated. On one memorable occasion in Australia, I would undoubtedly have been knocked unconscious had it not been for the diplomatic skills of our hostess.

It is rare to come across anybody who feels neutral or unconcerned about the issues surrounding the forbidden drugs. This is because you don't have to be a user to be profoundly affected by them on a personal level. Parents concerned about what their children are up to; teachers, doctors, lawyers, and other professionals who come into contact with drug use in the course of their work; victims of the spiralling levels of drug-related crime; policy makers, social activists, and libertarians; and almost anyone who ever opens a newspaper.

This sense of personal involvement introduces an emotional tone into any debate which can lead to bombast and speculation taking the place of a calm appraisal of the various dilemmas. 'Stamp out reality' was a typically provocative slogan of the hallucinatory Sixties, but it would also seem an apt motto for some of the modern rhetoricians and gurus who support or oppose the 'war on drugs.'

Recent opinion polls have indicated that many North Americans think that illegal drugs represent the nation's number one problem, with the related nightmares of AIDS and crime close behind. We don't yet seem quite so concerned in the United Kingdom, possibly because organized crime is less high-profile, and many of us still do not connect the soaring levels of muggings, car crime, and burglaries with someone's need to feed a drug habit. The bald truth is that tens of thousands of people in

Britain are faced with the challenge of raising £100 or so every day of their lives to maintain their supply of heroin or cocaine, with no legal income except a social security cheque worth about £40 a week.

Discovering that they are addicted is sometimes the first inkling that people have of the potency of the drugs they have been ingesting. Only the other day, this account was published in the letters section of a national newspaper:

'I made the decision to stop ... in my quest for a healthier lifestyle, and [had my last dose] ... last Thursday morning. By evening I had a mild headache. On Friday morning I had a fierce headache and felt dizzy and sick. I could not go to work. Friday evening I was physically sick and the headache was worse, with shooting pains in my neck and shoulders. Saturday I still had a bad headache, but I wasn't feeling sick and pain relief was possible.

Today, Sunday, I have a mild headache but I still feel drowsy. I hope I shall continue to feel better tomorrow.'

This writer was describing the consequences of breaking her addiction to coffee and tea. Many people today would either not classify caffeine as a real drug at all or would see it as being of negligible potency. My eleven-year-old daughter enjoys a cup of tea most days and nobody reports me to Social Services, yet in seventeenth-century Britain coffee was regarded very much as cannabis is nowadays: a potent, mind-altering chemical which should be treated with respect. Since one only becomes aware of being addicted to something when it ceases to be available, most people remain blissfully unaware that they are indeed hooked—just as many nineteenth-century Britons and North Americans remained unaware that they were hooked on opium or cocaine since these too were freely available at that time.

Cultural conditioning is very powerful and accounts for much of the glaring inconsistency in attitudes towards familiar and unfamiliar drugs, and it is fascinating to look back and see how these cultural attitudes have fluctuated over the decades. The public perception of cocaine is a case in point: from a harmless tonic and constituent of that refreshing beverage Coca-Cola in the nineteenth century to a terrifyingly addictive narcotic fit only for debauchees and criminals between the world wars, to being *de rigueur* for everyone who aspired to a Porsche or even just a mobile phone in the Thatcher/Reagan years. The properties of cocaine, although significantly influenced by the way it is used, have remained constant, but the way they are interpreted by 'experts' and the public-at-large continually changes.

My aim in writing this book has been to set out for the interested reader with no special experience or scientific knowledge what is known about the drugs themselves and why people take them, the meaning of addiction and the treatments that are available for it, and the natural history of drug use. I have also described the current policy on drugs in Britain and the USA, and discussed some ideas for reducing drug-related problems in the future. I have not tried too hard to resist expressing some personal opinions along the way and I hope these are consistent with the known facts. This will be for the reader to judge and react appropriately with assent or outrage; the only response that would disappoint would be indifference.

Oxford P.R.
June 1994

Contents

Part I

1
Why use drugs?

Most readers will be neither surprised nor disturbed to be reminded that almost every adult in Britain and North America uses drugs every day of their lives in the form of caffeine, alcohol, or tobacco. In sharp contrast to this equanimity, their reaction to the news that getting on for half of teenagers in the UK will have experimented with a street drug by the time they are sixteen is likely to be one of dismay and incredulity*. But as we shall see in the coming chapters, the simple truth is that a sizeable minority of the population is prepared to flout the law and face considerable risk and inconvenience in order to consume forbidden substances. Why do they do it?

This question can be approached on two levels: first, by contemplating the attraction or function of intoxication, and secondly, by a more utilitarian search for those human or environmental characteristics which can be shown to make initial experimentation or persistence with drugs more likely.

A number of writers interested in psychoactive drugs have come to the conclusion that a desire or need for the experience of 'altered consciousness' has always represented more to mankind than mere hedonistic self-indulgence. Andrew Weil (*The Natural Mind* 1973) has argued that it is a basic human appetite, recognizable throughout history from the most primitive beginnings and beyond. Of interest here is the observation that some animals in their natural habitat seem interested in intoxication, being prepared to experience it repeatedly. Dogs have

*Throughout this book, prevalence figures are based upon published surveys, seizure rates by police and customs, databases, and official audits. All of these sources have their shortcomings, and readers should take them as approximations rather than hard facts. Source documents are to be found in the bibliography.

been observed snuffling the fumes from rotting vegetation to the point of incoordination, and an acquaintance of the writer had a cat which confounded its carnivorous nature by gobbling up any marijuana she inadvertently left lying around.

Weil argues that drug-induced mental changes are part of a whole range of natural 'altered states' which fulfil an important revitalizing or insight-producing function. Other examples of these would include twilight states between sleep and waking, daydreams, trances, meditation, hypnosis, and delirium. He sees evidence of the in-built attraction and need for these states in the antics of children making themselves dizzy, faint, or even unconscious by whirling or breath-holding. He quotes a more famous writer, Aldous Huxley: '... the urge to escape, the longing to transcend themselves if only for a few minutes, is and always has been one of the principal appetites of the soul. Art and religion, carnivals and saturnalia, dancing and listening to oratory—all these have served, in H.G. Wells's phrase, as Doors in the Wall.'

The writings of Weil and others challenge the assumption that consciousness-alteration is undesirable, dangerous, or 'wrong'. Does 'reality', he asks, really equate to ordinary waking consciousness? Huxley certainly did not think so, and modern brain research has given support to his conception of the brain as a sort of filter or reducing-valve, diverting from conscious awareness all information that has no immediate survival value. He felt that the inhabitants of a modern materialistic society progressively stripped of the spiritual dimension had more need than ever to transport themselves beyond such sterile boundaries. The ubiquitous street graffito of the late 1960s I referred to in the *Preface* called on people to 'stamp out reality!'

This is a highly controversial line to pursue, because it suggests that those who use mind-altering drugs, who take the chance of freeing themselves from a perceptual strait-jacket, may reap significant benefits from the experience. At the very least, they may gain insight into the robotic, stereotyped nature of much human behaviour. This can be a profoundly unsettling realization, and none of these writers suggest that it is an undertaking without risk. Weil concedes that drugs can

'... hurt your body, hurt your mind, impede your development'. But the possibility does arise that these might be risks worth taking.

For the remainder of this chapter, I am going to concentrate on the influences that lead one person to experiment with drugs, and another to avoid them. Despite prohibitive laws and the activities of police and customs, illicit drugs remain highly accessible to anyone who wishes to obtain them. Not just the 'softer' drugs, but heroin and cocaine are available on the streets or in the pubs and clubs of every town in England. A survey of 14-year-olds in the north of the country showed that 59 per cent had been offered at least one illegal drug. Sixty per cent of those offered a drug (36 per cent of the 776 youngsters sampled) had tried it. The drugs used were cannabis (32 per cent of total sample), nitrite poppers (14 per cent), LSD (13 per cent), solvents (12 per cent), amphetamine (10 per cent), and Ecstasy (6 per cent). The gender ratio of users was equal, and this group of adolescents had virtually no interest at all in heroin or cocaine. These figures are quite similar to others reported elsewhere. Almost half of North America's high school seniors have experimented with illicit drugs at least once (US Department of Health Survey 1991).

It is interesting to note that more than 50 per cent of these experimenters who have decided to try a drug once will either not take it again, or use it only very occasionally. In the case of glue-sniffing, only a quarter of those who try it once will repeat the experience. One drug which seems not to be associated with this preponderance of once-only or very infrequent users is heroin. In one large survey, the number of adolescents who had tried heroin was small, only 1.7 per cent of the total sample. But of those who had used it once, 65 per cent used it again, and of these 68 per cent were using it weekly and 15 per cent daily. This drug, and crack cocaine too, are harder task-masters.

A large London survey showed that 20 per cent of 11–16 year olds had used street drugs at least once, but this may be an underestimate since the study did not include a sizeable group of truants. As expected, there was a steady increase of prevalence with age, with 11 per cent of 13-year-olds and 26 per cent of 16-year-olds reporting some experience. The biggest jump in prevalence in this and other surveys is at 13–14 years, suggesting that this is the most crucial age at which to target good drug education. This survey is consistent with most others in demonstrating that heroin, and less strikingly cocaine, are of interest to only a tiny number of teenagers.

A significant prevalence in younger adolescents is worrying because American research suggests that use of an illegal drug before the age of 15 years is associated with drug-related problems in later life. Drug use in children under 14 is commoner in girls than boys. Thereafter, experimentation is much commoner in boys, but the gender ratio is equal for those who progress to repeated drug use.

Peak age of drug use is between 18 and 21 years with a steady decline thereafter. This decline correlates with progressive involvement in the structures of society as work, marriage, and parenthood take their toll! Married people use less drugs than those who are single or divorced, but the highest rates of all occur in men living with women to whom they are not married. It is well known that in later life women are much higher consumers of prescribed psychoactive drugs, while men seem to rely more on alcohol to soothe, or inflame, their nerves.

If you ask teenagers why they use drugs, they will say that they like the experience, that it is pleasurable, that it takes away unpleasant feelings of shyness, anxiety, or lack of confidence, that their friends all do it, or that it makes them feel pleasantly rebellious and independent. The scientific papers by the dozen which have been devoted to the subject do not really take us much further than this. All that effort to confirm what one would intuitively expect! But the interesting question remains: what makes one teenager say ‘yes please’ and another ‘no thanks’?

Most of the research has focused on three dimensions: sociocultural influences, personal characteristics, and inter-personal factors. Although the first dimension is very important in the formation of dependency and drug-related problems and will be examined in detail in chapter 10, it seems to have relatively little influence on initial experimentation and subsequent occasional use. The fact that drugs are slightly more accessible in urban areas does not seem to have a discernable effect on frequency of experimentation in schools.

Turning to the contribution of personal characteristics, the possibility that genetic make-up may have a part to play is raised from time to time, though it is not a particularly fashionable idea at the time of writing. Although there is some human data which suggest that it may indeed affect susceptibility to excessive alcohol consumption, experimental evidence to indicate that this might hold true with other drugs is lacking at present. This may simply be a reflection of the many practical difficulties involved in carrying out such research. Animal experiments have demonstrated that genetic manipulation can be made to affect the response to drugs in ways which in humans would be likely to influence the pattern of future use. It is quite easy to breed animals which are more or less sensitive to specific drug effects, prone to tolerance, dependence, or withdrawal symptoms, or likely to develop drug-seeking behaviour. There is no obvious reason why these genetic traits should not evolve naturally in humans.

There are a number of personal attributes, attitudes, and behaviours in early adolescence which can be shown to predict, albeit weakly, experimentation with street drugs at some point in the future. On the behaviour side, these include a demonstrated fondness for beer, cigarettes, and wine; minor delinquency; early sexual experience; active participation in political or other 'protest activity' (American studies only!); and a general willingness to get involved in exciting but risky activities.

There have been some attempts to explain individual differences in primarily physical terms, for example linking willingness to take risks with blood levels of the hormone testosterone, but most effort has been concentrated on the psychological dimension. Unsuspecting adolescents in various university cities throughout the world have been exposed over the years to bundles of bulky questionnaires and hordes of earnest interviewers. Well over a hundred papers have featured the gargantuan Minnesota Multiphasic Personality Inventory.

Not surprisingly, all this effort has generated a host of associations with future drug use. For what it is worth, the typical drug-user-to-be is likely to possess at least some of the following characteristics: rebelliousness, nonconformity to conventional values, and a tolerant attitude towards unusual or deviant behaviour; a relative lack of ambition and commitment to school work or career building; independent mindedness, self-reliance, and a reluctance to abide by rules; impulsivity; preoccupation with pleasure-seeking and risk-taking; a history of physical or psychological illness, impaired emotional well-being, or consistently low self-esteem.

The only formal measure which consistently demonstrates a predictive ability is called the Sensation Seeking Scale. 'Sensation seeking' has been defined as 'the need for varied, novel and complex sensations and experiences and the willingness to take physical and social risks for the sake of such experience' (Zuckerman 1979). The Sensation Seeking Scale is a 40-item questionnaire which provides a global score and ratings on four sub-scales labelled: 'thrill and adventure seeking', 'experience seeking', 'disinhibition', and 'boredom susceptibility'. People who like cigarettes, alcohol, and stimulating foods also score high on this scale, which its author believes provides a measure of 'sensitivity to reinforcement'. He predicted that high scorers would prefer stimulants and hallucinogens to other classes of drugs, but this has not proved to be the case. It seems that high scorers value altered consciousness in whatever form it may take.

The Sensation Seeking Scale does not seem to me to have much value as an explanatory tool. It is like trying to account for the consumption of cakes by a measure of liking for sweet things. A person who enjoys excitement and doesn't dwell too much on the risks is more likely to race motor-bikes or smoke cannabis than someone who lacks these attributes. So what? The majority of people with the same attributes will not do either of these things, but will select any one of a thousand other pursuits which fulfil their requirements but do not happen to be on the investigator's list. The real question to be answered is one step back: what are the precursors of a craving for excitement and disdain for risk?

What it all boils down to is that most adolescents and students try a drug for one of three main reasons: fun and curiosity; to fit in with their friends; or to medicate themselves for some unpleasant emotion and forget their troubles. One investigator explored the proportion falling into each category, and found that fun and curiosity accounted for half, peer pressure a third, and self-medication only a fifth of the stated motivation. This pattern held true for all drug types except the stimulants; here, fun and curiosity was the prime mover for 40 per cent, with the other two categories 30 per cent each. These results confound the 'tension-reduction hypothesis' which proposes that the majority of drug-takers do so in order to relieve anxiety or unhappiness. This error is a product of basing assumptions upon the highly non-representational problem drug users who present to doctors and others for treatment. The study also revealed a reliability problem with the Sensation Seeking Scale; men scored higher than women, and whites more than blacks. This sort of distortion is most undesirable in questionnaires.

Not surprisingly, cannabis-using adolescents have positive views on the meaning of the drug experience, and have very different ideas about the possible risks than non-users. They tend to be more tolerant and *laissez-faire* in their attitude towards unpopular or ostracized groups, for example homosexuals or ethnic minorities. Political views are more likely to be left-of-centre. School or university performance tends not to glow, which may be related to a dearth of ambition. This may stem from a feeling of lack of personal control and autonomy in the path life is taking. If a true métier can be discovered, application and energy can confidently be expected to reappear. The seeming apathy may thus be an understandable reaction to circumstances rather than a drug effect or character trait.

The usual progression is to start with cigarettes, wine, beer, or solvents, then move on to cannabis. The greater the consumption of cannabis, the more the chance that other drugs will be tried. Stimulants, 'party drugs', and hallucinogens might follow, and heroin is the end-point for that tiny minority who have tried everything but are still searching. The only drug which not infrequently post-dates heroin is cocaine. It must be emphasized that there is no evidence that this is a *causal* sequence; use of a drug is not *caused* by exposure to the drug proximal to it in the sequence.

The third important dimension shaping amenability to drug use is the network of personal relationships which envelops every individual, and it makes sense to look first at the influence of the family. The children of an alcoholic father or a mother in receipt of a prescription for psychoactive drugs are more likely to use an illicit drug other than cannabis, and the theory put forward to explain this is that they have learnt from their parents that psychological stress requires a chemical solution. Consumption of illicit drugs is higher in teenagers who lack one or both of their natural parents, who come from families with high levels of stress and conflict, whose parents themselves use illegal drugs, or who have a close relative who is 'antisocial' or 'alcoholic'.

Drug-taking may simply be a small part of a more general questioning or challenging of established mores, and this pattern may well find its roots in families with 'permissive' or *laissez-faire* values. Conventional, conservative families operating along more authoritarian lines are likely to produce children with an inbuilt resistance to deviating from the norm, and less inclined to buck the system. Each style has its advantages and disadvantages. Although they sometimes come a cropper, we can all think of risk-takers who have hit life's jackpot in one way or another.

A person's friends and peers have an even more powerful immediate impact upon attitudes and behaviour than the immediate family, but this is a much more short-lived and circumscribed effect. Whereas the family still continues to exert its influence long after the person has left home and become independent, the power of the peer group fades rapidly as the person moves to a fresh milieu or social circle.

Not surprisingly, surveys of friendship groups show that they share attitude and belief systems, but it is uncertain how much this is due to like selecting like ('assortative pairing'), or one person influencing another. In other words, do similar people simply cluster together, or do certain individuals produce conformity by inducing changes in others. Influence can be direct through persuasion or role-modelling, or indirect by a subtler inculcation of beliefs and values. Indirect influence is much longer-lasting, and is the characteristic effect which the family exerts. Whilst immediate and overtly-stated parental rules may seem totally ineffective in curbing adolescent excesses, the subtler modelling of attitudes and behaviour results in psychological structures or 'mind-sets' which remain active for a lifetime. Unfortunately, they may not result in the outcome anticipated or intended by the parents. A father preoccupied with the virtue of hard work and long hours might expect his progeny to be impressed by the money and prestige this is producing; instead, the child may be taking note of the loneliness and lack of support experienced by the mother, and planning his own life accordingly.

The family also seems to play a vital role in determining an individual's self-esteem, which I would define as that sense of contentment and self-acceptance that stems from a person's appraisal of his own worth, significance, attractiveness, competence, and ability to satisfy his aspirations. Coopersmith (1967) has identified what he believes to be those characteristics of families which maximize the likelihood of a long-lasting high self-concept in their offspring: unconditional acceptance of children by the parents, foibles and all; clearly defined and enforced limits to the children's behaviour; respect and latitude for individual actions and decisions within these defined limits; and high self-esteem in the parents. As one might expect, having high self-esteem is enormously valuable quite apart from the obvious feel-good factor. Such people are likely to be more confident and self-determining, better able to handle stress and criticism from others, more willing to express a controversial opinion and resist pressure to conform, enjoy better relationships, and be less likely to suffer from anxiety and depression. Low self-esteem is a consistent finding in those who develop drug-related problems, but whether this is cause or effect is unclear. What is clear is that many of the personal characteristics which one might expect to be associated with a vulnerability to compulsive drug use have been linked with poor self-esteem: submissiveness, a tendency to depend upon others and constantly crave approval, lack of confidence, a sense of helplessness, anxiety, low expectations for the future, and a readiness to give up if the going gets tough.

You are much more likely to experiment with drugs if the peer group to which you aspire sanctions drug-taking, appears to use them without obvious problems, and gives favourable reports of the experience. The effect of this immediate circle of friends and acquaintances easily outweighs any conflicting pressure emanating from society at large and, albeit temporarily, from your family. Becoming warmly acquainted with established aficionados of any particular drug, be they beery medical students or pot-smoking jazz musicians, is by far the strongest predictor that you will try it yourself.

2
The consequences of drug use

A particular individual's reaction to a drug is shaped by many factors: the pharmacology of the drug, its purity and the presence or absence of active contaminants, the route by which it is taken, the way it is metabolized and stored by that person, and possibly by his or her unique pharmacological make-up before exposure to the drug; the individual's personality and experience, mood and attitude at the time, and expectations as to what will happen; the nature of the physical and social environment in which the experience happens. The interaction of so many variables, some of which are fleeting and unstable, means that the drug can have different effects on different people, and on the same person at different times and places. This will be discussed further in Chapter 10.

The great majority of people who use illegal drugs never come to the attention of doctors, lawyers, or policemen. They are invisible to research unless briefly flushed out of cover by population surveys, and so do not feature in the statistics of the outcome literature. The problem that this presents to those with an interest in understanding the consequences of drug use beyond the boundaries of the addiction clinics can be illustrated by reference to the recent cocaine epidemic in North America. Population surveys indicated that over 25 million American citizens had tried the drug in one particular year in the 1980s. Animal experiments, emergency room surveys, and the pronouncements of experts suggest that it is one of the most toxic and addictive of street drugs, yet only a tiny proportion of these people presented for treatment or help of any kind, or were ever prosecuted. The rest remained completely invisible, and we have no idea how things turned out for them. Did they just snort the stuff a few times, enjoy it or not as the case may be,

then simply drop it as new priorities took shape? Similarly in the case of heroin, of the one per cent or even two per cent of the general adult population who tell researchers that they have tried it at least once, very few go on to contribute to the casualty statistics.

The discrepancy between the numbers who admit to having at some time smoked cannabis, or swallowed LSD, Ecstasy, or amphetamine, and those that ever present for help with problems related to these drugs is even more enormous. One can only assume that the vast majority use these drugs only briefly or in small amounts, or that they prove pretty well harmless to most people.

Since next to nothing is known about the natural history of drug use in this invisible population of relatively problem-free users, one naturally looks with interest at the few studies that have been carried out. Every heroin or cocaine user in treatment can tell you about friends or acquaintances who seem able to limit their use of these drugs to weekends or special occasions, but it is all so subjective and anecdotal. Besides, who can be sure that these apparently controlled users are not simply in transit to addiction or abstinence?

A handful of American studies do give limited support to the idea that relatively stable, controlled use of opiates is possible. Such people often develop rituals and routines which ensure that the drug remains just a part of their lives, rather than its epicentre. We shall see in Chapter 10 that environmental and social context is all-important to the formation and maintenance of addictive behaviour. It seems that if the drug-taking can be ritualized and compartmentalized, then it can be more easily contained. The social drinker limits her consumption of alcohol to particular times and places, sanctioned and reinforced by society's rules. Studies among opiate 'chippers' suggest that many are successful in creating a similar structure around their heroin use. Such people inevitably have other strong interests and competing activities, clear plans for the future, a sense of being in control of their lives. They are wary of tolerance, that ever-decreasing pay-off from the drug consequent upon too-regular consumption, recognizing that this is the 'hook' which can draw them into dependency.

Many chippers do give a history of episodes of addiction, but usually seem to have been able to overcome these without professional help. Some oscillate between drug and 'straight' cultures, while others lead completely conventional lives apart from this particular divertissement. Intermittent users are more likely to prefer snorting or smoking heroin to injecting it.

In one large sample of young men selected because they stated that they had tried heroin at least once, only a third had ever reached the point of using it almost every day. Two-thirds of those who had progressed to monthly indulgence also reported periods of daily use. This suggests that it is difficult to avoid getting hooked if your use is anything other than quite infrequent, a possibility which is supported by observations on American soldiers who used heroin in Vietnam. Although the majority of these did not become addicted, 73 per cent of those who used more than five times did become dependent. This group also demonstrated that a period of addiction does not mean that future use will inevitably give rise to re-addiction; 36 per cent of the once-addicted sample did use heroin again in an infrequent, controlled manner on their return to the US.

Cocaine use has been examined using the 'snowballing' technique. Cocaine users identified through advertisements, personal contacts, or fieldworkers are asked to put the investigator into contact with similarly inclined friends and acquaintances, who in turn are asked to do the same thing. By this means, a sample is uncovered which does not have as its common denominator a connection with treatment services, and which should therefore be more representative of drug users in general. In groups of people discovered in this way dependence upon cocaine is the exception rather than the rule though, of course, one cannot be certain that these apparent copers are not merely in transition between control and addiction.

Six patterns of cocaine use have been described: an initial high intake slowly diminishing to zero; a slow and steady increase; stable consumption from the beginning to the present; escalation to high doses followed by a decline of similar rapidity;

regular intermittent use; irregular intermittent use. Progression to heavy use and back to controlled use or abstinence seems to occur in more than half the subjects, and a further third seem able to cope with regular or irregular controlled use of a fairly consistent amount over lengthy periods. Quite a proportion of these will be using small doses infrequently. Long-term prospective studies are very rare and involve only small numbers but, for what they are worth, also suggest that dependency upon cocaine requiring professional help is quite unusual. Most individuals who are able to take cocaine in a controlled manner over many years do seem to use it very sparingly, no more than once or twice a month. As we have seen, periods of heavy use do occur for some, and many of these people appear to take notice that the writing is on the wall. In one sample, only 20 per cent of those who had escalated to daily use were still taking cocaine one year later. On the other hand, many years of controlled use can sometimes lead to an unwarranted sense of invulnerability, since loss of control after a decade or more of problem-free experience is also well recognized. All of this research focuses upon snorted or, much more occasionally, injected cocaine and less is known about the consequences of flirting with crack, except that certain populations seem particularly vulnerable to compulsive patterns of use (see Chapter 10).

The conclusion from this work seems to be that addiction to heroin and cocaine occurs in only a minority of those who try them. Since, as we shall see, addiction usually has devastating consequences for those who do fall within its grip, it is particularly unfortunate that it remains so difficult to predict with any confidence who will be so afflicted.

Use of heroin, cocaine, and the other recreational drugs can interact with many aspects of ordinary living. Effects on energy, concentration, mood, or physical health can cause tension and disruption within families, and confrontations or disappointments at school or in the workplace. Many American employers now operate drug (and alcohol) policies which involve regular urine screens for illicit drugs. Testing positive in such screens can result in dismissal even if there has been no discernable effect on performance.

In Britain, the Road Traffic Act requires applicants or licence holders to inform the Driver and Vehicle Licensing Agency (DVLA) of any 'prescribed or prospective disability' which could pose 'a source of danger to the public'. Such disability includes 'addiction/use or dependency on illicit drugs' or 'chronic solvent misuse or dependency'. If the DVLA receives evidence of 'cannabis misuse' six months off driving will be required, and twelve months if abuse of other drugs is confirmed, including 'any psychoactive drug currently fashionable'. A professional driver faces revocation of his licence for at least three years, with strict criteria to be satisfied before it can be restored. It is an offence to be in charge of a vehicle whilst 'unfit to drive through drink or drugs'. Consultant-supervised maintenance on oral methadone (see Chapter 11) is no bar to retaining one's licence, but if injectable prescriptions of controlled drugs are provided it must be surrendered. Covert drug use revealed in testing following an accident may invalidate insurance cover.

Approximately a third of all North American drug users are women of child-bearing age, and more than half a million babies are exposed to illicit drugs *in utero* each year. The risks posed by the various drug groups are described in the relevant chapters. Damage to the fetus, though theoretically possible with most drug types, is extremely rare when the mother is an occasional or social user, though some adulterants commonly found in street drugs are potentially highly toxic. Drug users overall fall into the 'high normal' range for congenital abnormalities—approximately 3 per cent of live births. Common sense (but not much in the way of hard data) suggests that risks will be greater for addicted mothers who are likely to be consuming more drugs, and may also be suffering from the effects of poor nutrition, chronic infections, lack of family support and ante-natal care, and a chaotic life-style. It is perhaps worth noting that *all* drugs taken during pregnancy—legal, illegal, or prescribed—especially during the first three months, represent a potential risk to the fetus and should be avoided if possible. Tobacco is identified as a cause of low birth-weight and premature delivery, whilst a heavy consumption of alcohol has been associated with malformations, growth retardation, and impaired mental capacity.

Most people who use illicit drugs socially, and many who are dependent upon drugs, make as good a job of caring for their children as any other parents, in my experience. In the UK, however, the Children Act (1989) requires that in cases where a child appears to be at risk in terms of health or development, it is incumbent upon social workers and other professionals to ensure that the interests and welfare of that child take precedence over all other considerations. Interventions will almost always be aimed at helping the parent or parents to provide the emotional and physical environment that the child requires.

Long-term research following the progress of regular users of cannabis, the hallucinogens, or amphetamine is virtually non-existent. As far as one can judge, it appears that such drug use often peters out as middle age approaches, though no doubt many people quietly persist into their dotage. The outcome statistics that do exist mainly relate to heroin use. These must be interpreted with caution since they can only focus upon *visible* drug users. Invisible users are presumably running their lives more efficiently, otherwise they would have become visible. There are variables which make it difficult to generalize beyond any particular sample, such as the availability of barbiturates on the black market. These were commonplace in the 1960s and 1970s but are now rarely encountered, were peculiarly lethal, and accounted for at least 60 per cent of overdose deaths. Most long-term follow-up studies quoted today began in the early 1970s, so mortality figures may be misleadingly high. Again, injecting habits have demonstrably changed since the threat of HIV infection was first recognized in the mid-eighties.

A number of attempts have been made to follow the progress of opiate users over years rather than months. Addiction is seen in these studies to be a chronic relapsing and remitting process which carries a significant mortality from accidents, illness, suicide and, in the US, homicide. Despite this, the outcome is not as gloomy as generally thought since up to half the subjects are found to be consistently abstinent after ten years. Indeed, in those addicts that survive, the natural history of the condition with its unpredictable ups and downs seems slanted towards slow but steady recovery, which may lead to the beneficial effect of treatment being over-estimated.

If this is the broad pattern of opiate addiction, the details vary considerably from study to study. Most addicts become heroin-free by middle age, with a typical addiction 'career' lasting about ten years. The fluctuating course is marked by longer and longer intermissions of controlled use or abstinence. Notwithstanding this trend, between five and ten per cent of addicts are more than 45-years-old. Some may become dependent upon alcohol or other drugs as heroin loses its appeal.

Physical problems are mainly related to life-style, poor injecting technique, and adulterants in street drugs. Pharmaceutical opiates are remarkably non-toxic and there are many documented examples of doctors and others with access to pure drugs who have injected opiates over decades and still managed to lead stable and productive lives.

American follow-up studies usually make grimmer reading than British ones. A quarter of the sample is likely to have died over a 20-year period, and between a quarter and a half remain addicted for the whole time. Nevertheless, a benign natural history is still discernable for survivors. One respected investigator published figures on a sample of male addicts which he followed for 20 years (Table 1).

Treatment is but one of a number of factors which may

Table 1 Twenty-year follow-up of male addicts

	Time after first hospitalization (years)		
	5	10	18
stable abstinence	10%	23%	35%
dead	6%	11%	23%
active narcotic addiction	53%	41%	25%
uncertain status	31%	25%	17%

From: Vaillant, G.V. (1988). *British Journal of Addiction* **83**, 1147–57.

impact upon the profile or duration of these addiction careers. The common denominator of those treatments which do have some demonstrable impact seems to be their ability to introduce or enhance structure in people's lives. Natural events which achieve the same result, such as a new relationship or job, can be just as beneficial. Factors which most commonly lead people to seek treatment include health or legal problems, pressure from the family, and boredom with the whole rigmarole and risk of obtaining black market supplies.

Short-term treatment measures, though they may be a necessary preliminary or stepping-stone, have not been shown to exert any effect at all upon long-term outcome. Detoxification is a case in point, with some people undergoing hundreds of these procedures in the course of their addiction career. In one series, 100 addicts clocked up 770 detoxifications between them, and on only 3 per cent of occasions was this followed by a period of abstinence of a year or more.

When detoxification is linked with some continuing therapy, results are often better than they appear at first glance. Although 71 per cent of addicts in one study had sampled heroin again within six weeks of completing detoxification, many of these did not revert to dependent use and there was a steady increase in the abstinence rate over time. At six months, 45 per cent were off drugs and living in the community, whilst a further 14 per cent were using only on an occasional basis. Almost all lapses had occurred in the afternoon or evening, mostly in presence of other heroin users. Doses tended to be lower than was customary before the detoxification, with less injecting. Only a minority of the abstinent group reported any problems with alcohol, though this has been more prominent in other series. A quarter regularly smoked cannabis.

The greatest number of lapses happened in the first week after detoxification, and the results from other studies show that 80–90 per cent of full scale relapses will have occurred within the first few weeks or months. Reasons given for relapse include re-exposure to the drug culture and drug-using friends, recurrent craving, inter-personal conflicts, unpleasant mood states or environmental conditions, and reduction of staff support.

A fairly stable finding is that long-term outcome relates to the length of time spent in treatment, and if that duration is less than three months it is unlikely to be more effective than detoxification alone or no treatment at all. Of relevance to this observation is the finding that methadone maintenance programmes, in America at least, retain clients better than drug-free outpatient programmes or therapeutic communities. Nearly half of methadone maintenance clients remain in treatment for a year, compared to a quarter of those who enter therapeutic communities and 13 per cent of drug-free outpatients.

Predicting outcome for the individual addict is very difficult indeed. Having a serious pre-existing psychiatric illness is a definite disadvantage, and there are some other relatively weak indicators of an unfavourable prognosis: poor pre-addiction social stability, employment history, or educational attainment; heavy involvement in the drug culture, or having a drug-using partner; history of serious criminal activity; heavy alcohol use before or during treatment; an inability to maintain long-term relationships; and, with relevance mainly to the US, raised in a culture different to that of the parents. Interestingly, the severity of the addiction in terms of frequency of use or quantity consumed has little or no bearing on outcome.

Two factors stand out above all the others in determining a successful outcome. These are achieving a structured and rewarding life-style, and changing location after becoming abstinent.

Addicts who are successful in beating their habit are usually those who are able to find alternative sources of interest and gratification, which lead to new routines and preoccupations. These may be achieved in a variety of ways such as a new relationship, a job, or even an alternative, less damaging dependence. The latter could take the form of religious involvement, or regular attendance at a self-help group. There is no doubt that membership of Alcoholics Anonymous (AA) is strongly associated with prevention of relapse into alcoholism, and there is no reason to suppose that this would not also be the case with Narcotics Anonymous (NA). It remains uncertain whether this is cause or effect as it is entirely possible that the same factors which maintain remission also cause attendance at the meetings rather than the other way round. At the very least, AA and NA are very effective at breaking the unhealthy obsession with self through their emphasis on helping others, and provide an inspiring model of attainable abstinence to those who may have begun to abandon hope.

An occupation or career fulfils many functions beyond the obvious one of generating a supply of money. It provides a sense of belonging and purpose, contributes to self-esteem and confidence, permits an escape from the humdrum tasks of day-to-day living even though it may itself be quite routine and dull, introduces the worker to another circle of acquaintance, and perhaps most important, adds structure and shape to life. Unfortunately, at times of high unemployment even the most adaptive and able people can find it impossible to find work, but it seems that this is particularly true of addicts. A study of a group of New York addicts showed that by the time they were 40, they had spent only 20 per cent of their adult lives addicted but 80 per cent of it unemployed. It is tempting to ascribe their vulnerability to addiction at least in part to this evident inability to find any productive niche in life.

The other quite striking predictor of a good prognosis is a change of domicile after coming off drugs. According to the published opinions of ex-addicts themselves, this seems to have played the decisive role in maintaining abstinence in at least a quarter of cases. One investigation followed a group of 171 heroin addicts over a period of two decades. Abstinence occurred for 54 per cent of the time spent away from the home city and for only 12 per cent of the time spent living in it. Abstinence lasting a year or more was three times more likely to be achieved away from home. However, 81 per cent of abstinent addicts relapsed within one month of returning to their home city. It is not difficult to imagine why this should be so. Conditioning to environment and people, long-standing drug-using friends, family pressures and conflicts, familiarity with the drug scene and the nuances of scoring presumably all played their part.

To summarize our knowledge of the natural history of drug

use, very little is known beyond anecdote about the long-term outcome of cannabis, LSD, ecstasy, or amphetamine use. Most physically and mentally robust people with reasonable social stability who use heroin or cocaine a few times will not become addicted.

Those who do become addicted are likely to find that their lives become grossly disrupted, and their habit carries with it a frightening mortality rate. Spending at least three continuous months in treatment of any sort improves outcome, and the best chance of maintaining abstinence or controlled use of drugs is for former addicts to build lots of structure into their lives, and move away from the area in which they developed their addiction.

Part II

THE WEEK

3
Cannabis

A hardy but otherwise undistinguished weed has provided the world with one of its most remarkable drugs. For centuries the psychoactive properties of cannabis have been put to use in religious and social rituals. Primitive and sophisticated peoples alike have valued its power to ease the burden of reality and entrance the senses. Alongside this, it has an intriguing history as a remedy for many of life's most ubiquitous and irksome discomforts. The use of cannabis as a medicine is described in the Egyptian Ebers papyrus of the sixteenth century BC, at which time it was prized as a powerful painkiller.

In the modern world, no other prohibited drug has provoked such polarization between its defenders and detractors. In no other case is reason so swamped by rhetoric on both sides of the divide.

The psychoactive properties of the hardy annual now labelled *Cannabis sativa* were probably first recognized in Asia, and there is a suggestion that it was the first crop to be grown for reasons other than food production. The Assyrians made use of it in the eighth century BC, the Chinese were certainly cultivating it by the fourth century, and it is described in Indian religious writings in the second century. It found an important place in ancient Chinese and Greek pharmacopoeias. Ceremonial, recreational, and therapeutic use has continued uninterrupted throughout Asia, Africa, Arabia, and South and Central America down the centuries. Traces of cannabis have been identified in the remains of a young girl who died near Jerusalem in the fourth century AD. Scientists have deduced that she was in all probability given the drug to ease the pains of childbirth.

The other great asset of the plant, its fibrous stem so useful for rope-making and weaving, was utilized by the Romans who initiated the cultivation of the plant in Britain. By Tudor times, this cultivation had been expanded on a very large scale but still the demand for the fibre could not be satisfied. The early settlers in North America were encouraged to grow the plant at the beginning of the seventeenth century, and very soon this formed a considerable rural industry. Apart from the fibre, the seeds were a source of oil for fuel and other commercial applications, and found to be most nutritious for birds.

This tenacious relative of the nettle grows wild throughout the world. Its psychoactive product is available in the East in many forms varying in potency, refinement, and expense, and goes under many names, including bhang, charas, ganja, kif, dagga, kabak, and hashish. In these endemic areas it is used ceremonially, but also quite casually by ordinary folk to alleviate fatigue and boredom. It was introduced into the Carribean from Bengal, and the population took to it most readily. The psychoactive and herbal properties of cannabis were probably first introduced to Europeans on any scale about a thousand years ago by the Moorish marauders who were then ravaging Spain and Portugal.

The weed was known in the West as Indian Hemp until Linnaeus christened it *C. sativa* in 1753, and it only began to emerge here as a herbal remedy on any scale in the eighteenth century. It was not long before it began to be seen as a rival panacea to opium in Britain, Europe, and North America.

By the nineteenth century, cannabis had entered the orbit of Western mainstream medicine. The list of indications steadily became longer and longer. It was recognized as an effective pain-killer, anticonvulsant, sleep inducer, anti-nauseant, and appetite stimulant. It was thought to be useful in the treatment of lung disorders such as bronchitis and asthma, and in the alleviation of mental illnesses such as melancholia and mania. Queen Victoria was a grateful recipient on more than one occasion, although the nature of her indisposition goes discreetly unrecorded. The American Dispensatory of 1851 also suggested it might be useful for relief of gout, rheumatism, tetanus, and delirium tremens.

Recreational use does not seem to have been very widespread

in nineteenth century Britain and North America, but it became a firm favourite among artists and intellectuals in the bigger cities. Dr Jaques Moreau de Tours was introduced to hashish whilst touring North Africa, and on his return to Paris in the 1840s founded the *Club des Haschishchiens* at the Hotel Pimodan with such chums as Dumas, Balzac, Flaubert, and Baudelaire. The doses consumed by the members of this club were monstrous by modern western standards, and the effect upon their already fevered imaginations was florid indeed.

Herbal use and medical interest in cannabis seems to have declined towards the end of the nineteenth century, and its recreational use in the West became low-profile and 'down-market'. However, use by the poor and labouring classes in the British colonies, and the competing interests of the distilling industry, led the British government to instigate a massive scientific enquiry into cannabis use and its possible dangers in Asia. The Indian Hemp Commission produced its seven volume report in 1894. This very comprehensive survey revealed no convincing evidence of 'mental or moral injury' from moderate use of cannabis. Excessive use was no more likely to occur than was the case with alcohol, and seemed to be more or less confined to those with an established tendency to be idle or dissipated. These findings have been broadly confirmed by the various detailed reviews that have been commissioned at various times through the twentieth century. These will be referred to in more detail below.

Cannabis was outlawed in Britain in 1928 when the government ratified the 1925 Geneva Convention on the manufacture, sale, and movement of dangerous drugs. It did, however, remain available in pharmacies for use in psychiatric indications as late as 1973.

In North America, the drug was associated in the earliest years of the century with the poorest sections of the community, and the reefer (marijuana cigarette) was introduced into the country by wandering Mexican labourers. However, it soon began to catch on more widely, initially among the mainly black jazz musicians and their followers, and started to develop an extremely negative image in the newspapers of the deep

South. Lurid stories linking it with horrific violence, crime, and sexual debauchery became commonplace, and induced a truly hysterical reaction to the drug in some quarters. This seems to have been driven, at least in part, by the hate and fear inspired by the minority groups which were particularly associated with it—immigrants and blacks.

A rich folklore and grammatical idiom developed around the use of cannabis, which became known by such names as weed, tea, gage, loco-weed, and Mary-Jane. The reefer or 'joint' was sometimes referred to as a 'mezz' after the white jazz musician Milton Mezzrow who seems to have enjoyed access to material of particularly fine quality. After a highly eventful life, Mezzrow died in 1972 at the age of 73.

By 1930 the steady flow of scare stories, most of which appear to have been entirely fanciful, had led 16 states to ban cannabis and in the same year the Bureau of Narcotics was formed within the Treasury Department. The first Commissioner was Harry Anslinger, and he was to spend the next 30 years doing everything in his power to blacken the image of cannabis and all who used it or spoke up in its defence, with truly fanatical zeal. In 1937, the Marijuana Tax Act effectively outlawed the drug nationwide despite the opposition of a number of doctors and psychiatrists, as well as people concerned with civil liberty.

All the contemporary scientific investigations contradicted Anslinger's propaganda campaign, but were studiously ignored or their authors pilloried by the Bureau and the media. The scholarly La Guardia Report (1944), for example, was based upon a highly detailed review of the medical, psychological, and social aspects of marijuana use in New York City. The report concluded that there was no evidence to suggest that cannabis induces aggressive or antisocial behaviour, increases sexual crime, or significantly alters the personality of the user. The scientists could find no evidence of addiction, or mental or physical deterioration in their subjects. Then as now, sensational anecdote was preferred to painstaking enquiry.

Contrary to the Bureau's statistics, cannabis use had a low prevalence in the general population right up to 1960, even though the weed was growing wild and unrecognized in millions of gardens throughout the nation. To most English people it was almost unknown as a fun-drug until the 1950s when immigration from the Carribean increased greatly, and the incomers brought their cannabis with them (ganja in some form was said to be used by up to 70 per cent of the population of Jamaica at that time). Marijuana began to turn up in London folk and jazz clubs, and the first white teenager to be busted found himself in the dock in 1952.

In the 1960s, as everybody knows, cannabis exploded out of these narrow confines. It became integral to the developing hippy and psychedelic movements, but also found its way into student life and the homes of otherwise quite unrebellious and non-deviant people. By 1970, upwards of 25 million Americans had puffed on the weed, and there were 10 million regular smokers. In England 4 million had tried it, including a third of all university students. That characteristic aroma even became familiar in and around the conservative corridors of the London medical schools in those heady years.

Prevalence expanded steadily through the 1970s, perhaps reaching its peak toward the end of the decade. By this time, the figures from contemporary surveys are rather remarkable (but see footnote page 3). Fifty per cent of people aged 18–25 questioned in the US said they had tried it at some time; 43 million Americans had sampled it, and 16 million were regular smokers. Ten per cent of American high-school students were puffing on it daily. In Britain, 20 per cent of employed people between the ages of 20 and 40 had smoked it within the previous month. These were not criminals or deviants. Millions of ordinary people liked dope and used it regularly, but seemed to be able to combine this with an ordinary life-style. The level of demand had and has enormous economic implications for the countries where cannabis has always been an important cash crop. In the early 1980s, the annual export from the Lebanon alone far exceeded 2000 tonnes, a significant contribution to the national economy.

So concerned was the British Government that it commissioned Baroness Wootton to head an investigation. Her report

(1968) concluded: 'Having reviewed all the material available to us we find ourselves in agreement with the conclusions reached by the Indian Hemp Drugs Commission appointed by the Government of India (1893–94) and the New York Mayor's Committee on Marijuana (1944) that the long-term consumption of cannabis in moderation has no harmful effects'. She also found no evidence to support the escalation theory which suggests that cannabis is dangerous because it leads on inexorably to experimentation with more dangerous drugs. She proposed that cannabis should be separated from heroin in the eyes of the law, and that the drug should be available for research and for use within medicine.

Despite its carefully argued, unsensational style, the report received the sort of histrionically hostile press reaction that Anslinger would have delighted in, and was ignored by the Callaghan government. Other investigations have concurred with the findings of Wootton. The Canadian Le Dain Commission (1972) came to a similar conclusion. The British Royal College of Psychiatrists (1987) stated that '... on any objective reckoning, cannabis must at present get a cleaner bill of health than our legalised "recreational drugs"'.

Strong calls for decriminalization have arisen from time to time. In 1972, the American Presidential Commission on Marijuana and Drug Abuse recommended that the possession of a small amount of cannabis for personal use should no longer be a criminal offence, and in 1977 the Carter administration formally advocated legalizing the possession of up to 1 oz of cannabis. Public opinion oscillates according to the image the drug has in the media at the time; a Gallup Poll in 1983 revealed 53 per cent of Americans to be in favour of decriminalization, but this figure had declined to 27 per cent in 1986. Generally, politicians seem to have concluded that espousing the decriminalization of cannabis has no value as a vote-winner.

Whilst there are a number of varieties of *C. sativa* named after their geographical location (e.g. *C. indica*, *C. americana*), they are all essentially the same plant. Resin content and shape depend upon ambient conditions. If grown in peaty or heavy soil in a warm, wet climate, tall solid plants ideal for fibre extraction result. Plants grown in sandy soil in hot, dry surroundings are rich in resin, most of which is exuded from the flowering tops of female plants, coating nearby leaves and stalks. The plants are harvested between July and September. A particularly potent form known as *sinsemilla* (literally, without seeds) is obtained by culling out the male plants before pollination can occur. This results in larger flowering heads in the female plants, and a particularly abundant yield of resin.

More than 400 chemicals have been identified in this resin, including at least 60 psychoactive compounds (cannabinoids). By far the most important active ingredient is called delta-9-tetrahydrocannabinol (THC) which was isolated in 1966. Cannabis is available on the black market in many different forms which vary considerably in their THC content. These include herbal material (often called marijuana or grass), resin compressed into blocks (hash), or less commonly a thick oil or tincture. As a rough guide, dried herbal material (grass) contains anything from 1–10 per cent THC by weight, resin around 10–15 per cent, and oil 15–30 per cent.

An average marijuana reefer or 'joint' contains around 300–400 mg of herbal material. The quality of the street supply and the consumer's smoking technique will have a profound effect on the amount of drug absorbed, but even the most expert smoker cannot achieve more than 50 per cent absorption; 25 per cent would be a more likely figure. The dose of THC delivered by a single reefer could be as little as 1 mg or as much as 30 mg depending on the quality and type of material. Since the threshold dose for intoxication is around 2 mg, the effects will be rapidly apparent from a single cigarette. The placebo effect is not unimportant in cannabis smoking, and is more prominent in naive smokers or those with suggestible personalities. However, most of the effects of cannabis can be shown to be dose-related to some extent.

The effects of smoking come on within a few minutes, peak around 30 minutes, and last for three to four hours. If you eat it, onset is delayed for one or two hours depending on whether the stomach is empty or not, but the effects last much longer.

The main effects are upon mood, attention and memory, perception, and patterns of thinking. These psychological effects are quite variable, and are described below. Heart-rate is increased, clumsiness and slurring of speech may be noted, and the eyes become reddened, but the physical effects of the drug are generally very mild. Body temperature may be slightly reduced. Fatal overdose due to cannabis alone has never been reliably reported.

In animals, the release of reproductive and thyroid hormones is suppressed, as is the immune response, but the significance of these findings for humans is unclear. THC seems capable of lowering the sperm count in men, but this may be a temporary effect. There are some reports that heavy cannabis consumption in pregnancy may be associated with low birth-weight and prematurity, and animal studies suggest an increase in still-births. Although none of this amounts to clear evidence of harm to the unborn child, pregnant women would be wise to avoid this and all other unnecessary drugs.

People intoxicated by cannabis are more prone to accidents or miscalculations. Reaction time, depth estimation, tracking a moving object, time sense, and recovery from glare are all impaired. The implications for car-driving or other coordinated activities are obvious.

The drug is metabolized within the body to both active and inactive metabolites, some of which are absorbed into fat stores and take a very long time to get rid of, so that urine tests for cannabinoids can remain positive for many weeks after a single exposure. It is theoretically possible for passive inhalers of other people's cannabis fumes to test positive if sensitive immunoassay techniques are used. THC is capable of crossing the placenta into the unborn child, and is detectable in the breast milk of smokers.

How cannabis works in the brain remains something of a mystery. Recent research suggests the existence of a specific receptor for the drug distributed quite widely in the brain, with high concentrations in areas known to be associated with emotion, memory, and reasoning. This suggests that the body produces its own natural cannabinoid, analagous to the 'brain's own opiates', the endorphins. The drug is known to impair short-term memory, and this is consistent with the high density of receptors found in the hippocampus, a brain structure particularly associated with learning and the coding of sensory information.

Cannabis is demonstrably a powerful pain-killer in animals, again by an unknown mechanism. Although this effect is definitely not mediated through opiate receptors, cannabis greatly reduces opiate withdrawal symptoms in animals. This attribute has not gone unnoticed by opiate addicts who are struggling with the pangs associated with an interrupted supply of heroin. In monkeys, aggression is reduced but so is the motivation to perform complex tasks. The drug reduces susceptibility to fits.

As already mentioned, cannabis was widely used by Western doctors in the nineteenth century, but this is not simply an historical oddity. Undeterred by its decades in the wilderness, there are modern-day physicians and psychiatrists who argue that THC or its derivatives have an important role to play in therapeutics. The indications that have been put forward, and which have support from animal studies or anecdotal reports from patients, include the reduction of sickness and nausea in people receiving cancer chemotherapy (the rigorous American Food and Drug Administration approved this use in 1985); pain relief in terminal illness, or for muscle spasm in multiple sclerosis (MS); epilepsy; glaucoma (because it reduces pressure within the eye); and migraine.

Of course, the law in both Britain and the US forbids such use. Until recently, 'compassionate reefers' could be prescribed to named patients in the US, but this humane practice has now been suspended. In the UK, law-abiding but desperate MS sufferers are driven to the black market because they find no other drug so effective in relieving their muscle cramps and relaxing sphincters. More speculatively, there is some evidence that cannabis has antibacterial and antifungal activity. Some psychiatrists believe it has potential in the treatment of anorexia nervosa, withdrawal syndromes from opiate and other drugs, and a range of psychological and behavioural abnormalities.

There are estimated to be around 300 million recreational cannabis users in the world today. Approximately 60 million North Americans (a third of those aged 12 years and above) and somewhere between 5 and 10 million British people are thought to have smoked it at some time. It has been pointed out that this considerably exceeds the numbers who watch football matches, or go to church on Sunday.

It is thought that somewhere in the region of 29 million American citizens are current users, of whom as many as 7 million smoke it daily. Americans spent in excess of $40 billion on the drug in 1984. In Britain today, a conservative estimate is that at least one in five young men and one in ten young women have tried it, and that half of these will have used it seven or more times within the last 14 months (Institute for the Study of Drug Dependency, 1992).

Cannabis is the target of 85 per cent of all drug seizures by the British police and customs authorities, and the number of seizures continues to rise steadily year by year. The figure was 59 400 in 1991, 10 per cent up from 1990. The quantity seized was the highest ever in 1989 at almost 60 tonnes, dropped back considerably for a couple of years, then in 1992 surged back once again almost to the level of the late-eighties peak, and continued an upward trend into 1993. The vast majority of this captured material is in the form of cannabis resin. The biggest single UK seizure so far occurred in 1991, with the discovery of 18 tonnes on an oil-rig supply vessel.

Well over 80 per cent of all those arrested for drug offences in Britain are charged with possession of cannabis for personal use; this amounted to some 40 000 people in 1991. Despite an increased willingness to caution such individuals by many police officers, more than a thousand people went to prison for possession of cannabis in Britain in that year.

Current sources of cannabis for the British black market include Nigeria, Jamaica, Ghana, Morocco, Pakistan, and Afghanistan. Lebanese and Nepalese hashish is now rare but highly prized, as are the famed 'Thai sticks' (high quality herbal material tied into tight reefers). Average street prices for the small-scale customer have remained remarkably constant over the years; around £25 for a quarter-ounce of hash, more for good quality marijuana. It is usually bought as chunks of resin or bundles of herbal material, but occasionally appears as powder, sticks, tablets, or hash oil. Rough and ready home-produced 'grass' generally consists of a *mélange* of shredded leaves, seeds, and stalks and is rather low in THC, but is usually a good deal cheaper. Current street names include blow, puff, spliff, draw, and weed.

Cannabis is usually less adulterated than most other street drugs, but may contain other inactive or active (e.g. datura-containing) plant material. Increasingly, it appears that poor quality cannabis may be impregnated with the much more dangerous phencyclidine (see Chapter 5) to increase its impact. Cannabis grown in the US or imported from Mexico or Central America may be contaminated with paraquat or other highly toxic herbicides sprayed from aeroplanes as part of eradication programmes.

Most users in the West smoke the drug in reefer cigarettes (joints) or more ostentatiously in clay 'chillums' or water-pipes. One can also improvize in many ways using perforated milk-bottle tops and other every-day materials. It is also sometimes brewed up as a tea, or used in cooking. When consumed by mouth, it is more difficult to measure the dose and people may get considerably higher than they had intended. The onset of activity by this route may seem particularly sudden and powerful.

There are hundreds of descriptions in print of what it is like to take cannabis. Although these accounts are rather variable, the discernable common theme is the essential light-heartedness of the experience. This is not, for most people, a serious or heavy drug. Its effect has been described as whimsical in nature, and the high it produces labelled, presumably by a non-enthusiast, as 'fatuous euphoria'.

The effect of any drug is greatly influenced by a range of factors quite apart from its pharmacology, and this is particularly true of cannabis. The personality, mood, and expectations of the user; the quality of the drug and the amount actually absorbed into the body; the nature of the environment, and the attitude and behaviour of other people who are around at the time will all shape the experience quite fundamentally. Some have argued that cannabis intoxication is a 'learned behaviour', that one has somehow to discover how to recognize and welcome the effects, and label them in one's mind as enjoyable. Others are unconvinced by this thesis, feeling it may be based upon a lack of first-hand experience, or hash of inadequate quality!

Smoking dope is usually a sociable activity. A joint, usually containing a mixture of cannabis and tobacco, is rolled from one or more cigarette papers to achieve the desired dimensions, and then handed round until it is nothing more than a tiny glowing stub (roach), highly prized for its heavy potency. This may get so hot and wizened that it requires a special holder (roach-clip). The experienced cannabis smoker inhales deeply and retains the smoke in the lungs for as long as possible to maximize absorption. This induces a peculiarly strangulated conversational style which may enrage the non-smoking listener whose patience is already sorely tried by the antics of the group.

After a few minutes of smoking, a sense of calm, contentment, and well-being is noted, often accompanied by a peculiar expanding or tight sensation in the head that users find hard to describe. There is a growing feeling of hilarity and sense of the absurd which transforms the most mundane or banal observation or witticism into the funniest thing in the world. There is a sense of good-will towards all mankind, a strong sense of optimism, and enhanced self-esteem. Inhibitions are reduced or dispelled entirely. The group may be seized with paroxysms of laughter or helpless giggling which can be most unappealing to the non-intoxicated observer.

In moderate doses, the effect is to enhance perception rather than distort it. Music, food, and sex, for example, seem more pleasurable and intense than usual but the user remains firmly in touch with reality. Time seems to crawl by very slowly, but the consumer feels active and talkative. With slightly larger doses, there may be a more pronounced sedative effect.

As time passes, enhanced appetite may become evident, and

many people feel a particular desire for sweet things ('the munchies'). A rather dry mouth generally requires a steady intake of fluids. Sudden feelings of depersonalization or unreality sometimes come on in waves, and the degree of intoxication often seems to ebb and flow.

Novices who have misjudged the dose may occasionally find themselves physically transfixed. Their thoughts are racing and they may desperately wish to say something or get up and move about, but are effectively frozen to the spot. On the other hand, more experienced users claim to be able to master their intoxication if circumstances demand it. One chap remembered an occasion on which he was very high indeed, when suddenly the doorbell rang. Giggling inanely to himself, he went to answer it and there stood a policeman in uniform. The constable had been passing and had noticed a car parked with its headlights on in the drive. Being public-spirited, he had decided to alert the householder and prevent the inconvenience of a flat battery in the morning. Our subject claims he was able immediately to master his hysteria, find the keys to the car and nonchalantly stroll over to it with the officer, engaging him in small talk and not arousing in any way the suspicion that this might be a drug fiend in the grip of his habit. He was uncomfortably aware of the sounds of jollity that continued unabated from within the house. When he rejoined his friends, he relapsed instantly into the prevailing merriment which was no doubt enhanced by a shared sense of relief that the party had not been 'busted'.

To the user, this sort of intoxication seems valuable and life-enhancing. Musicians may be convinced that their playing is more free-flowing and inspired, talkers that they are more witty and interesting, lovers that their sexual powers and enjoyment are augmented, writers that their imagination is broadened. Sceptics or 'straights' would say that any idea of improved creativity is pure illusion, literally a pipe-dream. Users are banal, clumsy, infantile, irritating and illogical, and almost entirely lacking in judgement.

Hangover from moderate use of cannabis is usually mild, but it is easy to over-sleep the following morning, occasionally by several hours. Larger doses may accentuate the sedative possibilities of the drug, or result in sensory distortion or frank hallucinations. Members of the *Club des Haschishchiens* referred to earlier in this chapter took huge doses by today's standards, and Gautier has left an interesting description of the phenomenon known as *synaesthesia*, where the senses become interchanged. Describing a vase of flowers, he wrote: 'my hearing became prodigiously acute. I actually listened to the sounds of the colours. From their blues, greens and yellows there reached me sound waves of perfect distinctness'.

The large majority of unwanted effects reported after cannabis use are minor in nature and brief in duration. Anxiety, occasionally building up to full blown panic attacks, may result from misinterpreting the rapid heartbeat which usually accompanies use of the drug as evidence of an impending heart-attack, or the psychological effects as a precursor to complete loss of control or madness. Dizziness or depersonalization may similarly be assumed to herald impending catastrophe. These symptoms are commoner in less experienced smokers who may also be more vulnerable to transient mood disturbances. Cannabis tends to heighten whatever mood was dominant before exposure; depressed or anxious people may well feel worse.

Impaired co-ordination and lack of judgement may prove to be a lethal combination for drivers, pilots, or others performing tasks requiring skill and prudence.

There are thousands of published reports of more serious adverse effects of cannabis, but scientific validity is in short supply. This is because of the many difficulties inherent in this sort of research, which include the need to rely on uncorroborated retrospective accounts of illegal activities by the subjects under study; lack of information on physical, psychological, or personality problems that may have existed before the subject began to use drugs; covert continuation of cannabis use or undisclosed use of other drugs currently or in the past; variations in the quality of illicit drugs, and the amounts absorbed; lifestyle factors such as poverty and deprivation, or unrecognized physical or mental illness. The scientific design of the studies is usually unsatisfactory, with inappropriate tests, inadequate statistics, or overinterpretation of the results. The prior assumptions of the experimenter may shape the design of the study or the interpretation of the findings.

One side-effect which is a genuine cause for concern is the effect of cannabis on memory, in particular the ability to learn new information. It is quite evident that memory can be acutely impaired in people who are actually intoxicated. A more important question is whether memory remains impaired for weeks, months, or even years after a person has stopped using the drug. This dilemma remains unresolved. One study does suggest that memory may remain abnormal in humans after several weeks of abstinence, but it has scientific limitations and requires confirmation. This is an area which cries out for high-quality research.

Heavy smokers may feel 'paranoid' from time to time, that is they imagine they are being watched or followed, or that everyone is against them. Such people will usually be able to accept that these feelings are just in their imagination, and they soon disappear if the drug is stopped or cut right down. It is possible to take such large doses, especially by the oral route, that a state of delirium may be induced with confusion, hallucinations, and delusions. Relatively unusual in the West, such reactions are reported more frequently in Asia and the Carribean. Occasionally, weird frozen postures may be taken up in such cases: a student who, fearing detection at an airport, swallowed a large chunk of hash was found some hours later by his flatmates half-reclining on the floor with his limbs held rigidly in an extraordinarily uncomfortable-looking and contorted way. He was fully conscious and seemingly undistressed, though bemused, but remained unmoving in this position for many hours. Within a day, he was completely back to normal, albeit somewhat sheepish.

A highly controversial question is whether cannabis is capable of producing a full-blown mental illness with hallucinations and delusions in someone who was previously perfectly healthy. Many papers have appeared in the medical literature which seem to suggest that this may be so, but the scientific shortcomings summarized above undermine their persuasiveness.

Although it is undoubtedly possible to develop a short-lived illness of this sort as a toxic reaction to high doses, on balance there is no convincing evidence that it causes long-term psychiatric illness *de novo*. People consuming large doses over long periods may continue to suffer psychotic symptoms as a form of prolonged toxic reaction, and this phenomenon may explain at least in part the many anecdotal reports of 'cannabis psychosis' which emanate from the East. Such formal surveys as exist in the Western world suggest that the prevalence of serious mental illness is no more common in regular, moderate smokers than in the general population. However, though it seems reasonable to accept, on present evidence, that 'cannabis psychosis' persisting for weeks, months, or longer after complete cessation of all illicit drug use is in reality schizophrenia, it remains an open question as to whether heavy use of cannabis may be one of the risk factors for the development of this condition. The drug certainly seems capable of provoking relapse or a worsening of symptoms in psychologically vulnerable people or those already in the grip of a mental illness. On the other hand, some people with chronic mental illness have claimed that the drug calms them, or reduces the intensity or intrusiveness of such distressing experiences as hearing voices when there is nobody there, or imagining that the TV is broadcasting messages to them personally. On balance, most doctors would recommend that people with a serious mental disorder or marked emotional instability would be wise to avoid cannabis.

Another high-profile problem that has been laid at the weed's door is that it induces a state of chronic apathy, passivity, and indolence known as the 'amotivational syndrome'. This reputation goes back a long way, and certainly finds its way into the pages of the Hemp Commission report as a cause of concern in nineteenth century India. In interpreting these anecdotal reports, one must bear in mind that the subjects described tended to be marginalized, unhealthy people perhaps suffering from malnutrition and almost entirely lacking in prospects, who were using potent cannabis in very large doses over long periods. There was some indication that people who were 'mentally unstable' to start with were more likely to use cannabis excessively; in other words, that mental abnormality was possibly a cause of drug use rather than a consequence. There is no evidence from Western studies to support the existence of amotivational syndrome except as a state of chronic intoxication in heavy users (so-called 'pot-heads').

Psychological dependency upon cannabis does occur, but is very uncommon. Heavy or compulsive users of cannabis may well have a history of emotional or mental disorders, and are statistically much more likely to be using other illicit drugs such as amphetamines, cocaine, or hallucinogens. Smoking cannabis sometimes induces flashbacks in LSD users. Physical dependency has not been reported, though abrupt withdrawal in long-term heavy users may result in irritability, anxiety, restlessness, and disturbed sleep for a few days.

Although the drug has a potentially useful airway-dilating effect in asthma, cannabis smoke is highly irritant to the lungs and contains numerous carcinogens. Presumably, heavy long-term smokers run the risk of developing bronchitis and lung cancer. Indeed, a major risk of smoking cannabis is that it may lead one to become a tobacco addict.

Part of cannabis's folklore is that it provides a great stimulus to crime, violent and otherwise. Although cannabis use is highly prevalent in prison communities and among the criminal classes, there is absolutely no evidence to suggest that it has a causal role. Cannabis-smoking school students do not commit more crime than non-smokers, for example. There is a much more compelling case for alcohol to answer in this dimension. Some prison officers express the belief off-the-record that prison society would be immeasurably more violent and disrupted if cannabis ceased to be freely available.

The animal data referred to earlier suggest that although direct evidence of damage to the human fetus is lacking, pregnant and breast-feeding women would be wise to avoid this and all other unnecessary drugs. Since it is known that cannabis freely crosses the placenta and takes a long time to metabolize, it is to be expected that babies born to mothers who smoked heavily towards the end of their pregnancy may be a bit lethargic for a while.

A long list of other terrifying consequences of cannabis use have been fielded, including shrinkage of the brain, chromosome damage, deleterious effects on reproduction and the immune system, and so forth. The evidence to support them is not convincing from the scientific point of view, but one cannot totally discount the possibility that real dangers lurk as yet undiscovered.

As things stand at present, the simple truth is that none of the exhaustive reviews of the available scientific data carried out from time to time at the instigation of various governments have revealed any convincing evidence that light to moderate use of cannabis does any harm whatsoever.

4
The stimulants

Cocaine

Chewing coca leaves has been part of everyday life in many South American cultures for several thousand years. For millions of people, the habit continues to fulfil a comparable role to that of coffee or tobacco elsewhere in the world. A wad of leaves tucked into the cheek for several hours, with perhaps a little wood-ash mixed in to increase salivation and enhance absorption of the active ingredients, will produce a sense of well-being, reduced hunger, and increased endurance; highly adaptive for a peasant in the mountains of Peru.

News of this intriguing plant first reached Europe in the sixteenth century, but it was not until the middle of the last century that its allure became widely publicized through the writings of an Italian physician who had witnessed its use amongst the Indians of Peru. The main active constituent, christened cocaine, was isolated in 1859 and soon began to appear in an ever-widening array of patent medicines and tonics. A famous example was Vin Mariani, containing the relatively modest dose of 8 mg per glass, which was available in France from the 1860s. This was soon extremely popular at all levels of European society, and reached the US in 1885. The medical response had at first been muted but interest grew rapidly, with articles by some famous physicians extolling the properties of the drug as little short of miraculous for a variety of ailments. Sigmund Freud was one of the many professionals swept along in the rush. In 1884, his great friend, the scientist Ernst von Fleischl, received morphine for the relief of severe post-operative pain and became addicted. This seemed an excellent opportunity to try out the wonder-drug. Freud got

hold of some to treat his friend, and was so encouraged by the initial results that he was inspired to write a review article which was, in his own words, 'a song of praise to this magical substance'. A frequently quoted advertisement of 1885 described coca as 'a drug which through its stimulant properties, can supply the place of food, make the coward brave, the silent eloquent, free the victims of alcohol and opium habit from their bondage, and, as an anaesthetic render the sufferer insensitive to pain . . .'. Coca-Cola, containing a few milligrams of cocaine in each glassful, was introduced in 1886 and marketed as a refreshing and stimulating alternative to alcohol.

In Britain, a growing recognition that the miracle drug could have a dark underbelly, reinforced by an element of professional expansionism, resulted in an abrupt restriction of outlets for cocaine alongside the opiates under the Pharmacy Act (1868). In the US, more and more people had begun to sniff or inject cocaine rather than imbibe it in drinks, and public awareness of the possibility of addiction began to grow in the last quarter of the century. The prominent American physician, Halsted, became a compulsive user as a result of his explorations into its local anaesthetic properties. By 1887, even such uncritical adherents as Freud had begun to experience a downturn of enthusiasm. His friend von Fleischl had become pitifully enslaved by the miracle drug, injecting well over a gram of it every day and able to think of little else. Commercial acknowledgement of this sea-change in public opinion was sign-posted when caffeine replaced cocaine in Coca-Cola in 1903.

Increasing legal restriction at state level in the US heralded the appearance of illicit sources of supply. It is interesting that the cost of cocaine on the streets of New York in 1907 was almost identical in relative terms to that in 1985. The Harrison Act (1914) limited the use of cocaine to medical prescription nationwide. In 1919, the Supreme Court ruled that maintenance prescribing to addicts was not a part of legitimate medical practice, and was thenceforth outlawed.

The black market in Britain was much less developed, but there were some widely published scandals involving cocaine in the early years of the century. During the First World War, concern grew over the sale of cocaine to soldiers on leave in London, and in 1916 Regulation 40 of the Defence of the Realm Act proscribed the supply of cocaine and a number of other drugs to members of the armed forces except on prescription from a doctor. With some additions, this Regulation was transferred to the Civil Law in 1920 as the Dangerous Drugs Act.

Cocaine underwent a progressive decline through the second quarter of the twentieth century in both Britain and North America, use being largely restricted to marginal groups in society, reaching a nadir in the early 1950s. It began a come-back in the middle 1960s, and the public and professional perception of the drug gradually came to resemble that benign and complacent one of a century before. Rapid expansion of use followed in the US, peaking around 1979. At that time, surveys indicated that 10 per cent of North Americans between the ages of 18 and 25 had used cocaine within the last month. By 1986, it was estimated that 40 per cent of Americans aged 25 to 30 years had tried it, and some three million people were regular users. The drug's social impact was greatly aggravated by the appearance of smokeable cocaine (crack) in 1985. The epidemic subsequently spread widely into mainland Europe but, although increases in cocaine and crack use have been noted, it has not afflicted Britain to a comparable degree at the time of writing, although ominous tremors are being felt in the form of rapidly escalating seizure rates.

The slopes of the Andes and the vastness of the Amazon basin play host to the two major species of coca-yielding shrubs. The cocaine epidemic in the developed world has created a black economy of dramatic proportions. Peru, for example, produced a quarter of a million tonnes of coca leaf in 1987, compared to 11 000 tonnes in 1959. Peru and Bolivia both retain a legal coca market at the time of writing, but the vast majority of production is destined for the black market, where the price to farmers greatly outstrips the official rate. The trade in cocaine was worth US $3.5 billion to Colombia in 1991. The effects on society of the titanic criminal organizations which have grown up around this industry are devastating.

Coca leaf is converted in the country of origin to coca paste, containing up to 80 per cent of the active ingredient. The pattern of cocaine use in these countries has changed markedly as this industry has developed. Paste smoking, virtually unknown before 1970, is now widespread and associated with much physical and psychological harm. The snorting of cocaine salt is on the increase amongst the more sophisticated users, but has yet to catch on in a big way.

The paste is further purified and refined to produce cocaine hydrochloride, a water-soluble powder which is very well absorbed by mouth and through the membranes of the nose. However, when the drug is swallowed the liver breaks down more than half of the dose before it can reach the brain. Rapid metabolism in the liver, blood, and elsewhere results in a halving of the blood level every 60–90 minutes. When the powder is sniffed ('snorted') absorption may be delayed because the drug narrows the blood vessels in the nose, reducing the flow of blood. The powder can also be dissolved in water and then injected into a vein. This results in much higher and more rapid peak concentrations in the blood, and cocaine hits the brain within 16 seconds of the injection.

Cocaine powder must be heated to more than 200 °C in order to vapourize it. At this temperature, most of the active ingredient is destroyed, the result of which is a disappointed smoker. Cocaine stripped of its hydrochloride salt (freebase, crack), however, vapourizes at a much lower temperature, so that more of the active ingredient remains intact. When smoked in a pipe, this material delivers drug to the brain at least as rapidly as would be achieved by injection, albeit at slightly lower peak concentration. A cocaine smoker can be distinguished from a snorter by the presence in the urine of a particular metabolite: methylecgonidine.

Frequent use by any route quickly results in reduced sensitivity to the euphoric effects of cocaine, but not to all of the unwanted effects (see below).

Cocaine stimulates the brain mainly by increasing the activity of the chemical messengers noradrenaline, serotonin, and dopamine. Its effectiveness as a local anaesthetic derives from its ability to block the initiation and conduction of impulses along nerves.

The effect upon one particular bundle of dopamine-containing nerves seems central to the drug's ability to produce a 'high'. It seems that this bundle of nerves may play a vital role in mediating the experience of pleasure in everyday life. Animals find electrical stimulation of the bundle very enjoyable indeed, and will perform any number of tasks to obtain this reward. Dopamine concentration in the bundle has been shown to be raised during exposure to such rewards as food or sex. Many of the drugs to which people can become addicted have now been shown to stimulate these nerves, but none more powerfully than cocaine (or amphetamine). These drugs rev the bundle up much more than any of the gratifications or delights life normally offers. This may account for the fact that animals, when given the choice of either cocaine or food (or sex, or warmth) will go for the cocaine every time. They will press a bar thousands of times to obtain a tiny dose. And given unlimited access, they will continue to dose themselves until they finally die of exhaustion.

Cocaine also stimulates the part of the nervous system nicknamed the 'fight or flight' mechanism (sympathetic nervous system). This produces an increase in heart-rate and systolic blood pressure, widening of blood vessels to muscles and narrowing of those to skin, a toning-up of the body's chemical systems generally, and an increase in the amount of oxygen entering through the lungs. The body is prepared for action.

Cocaine was used by doctors as a local anaesthetic for decades, but it has now been superseded by safer alternatives such as lignocaine and bupivacaine. Apart from very occasional use in anaesthetic eye-drops, no other medical indications remain.

Cocaine seizures in Britain increased rapidly in the 1980s, from 500 in 1981 to a peak of nearly 2100 in 1989, falling back slightly in 1990, and returning to 2000 in 1991. However, the quantity of cocaine seized has increased very remarkably from 600 kg in 1990, 1100 in 1991, to 2250 kg in 1992. Average purity lately has been around 50 per cent, rather lower than is customary. Crack seizures have increased sharply in the last few years, but remain relatively modest in quantity (1000 kg in 1991). Out of a total of 540 000 people arrested for drug offences in 1991, only 840 were charged primarily in relation to cocaine, a similar proportion to that of 1990.

Actual prevalence figures are hard to come by. In 1985, between two and three per cent of young adults questioned in surveys said they had used cocaine at some time, but another study indicated that many of these would be infrequent or once only users (see footnote page 3). Cocaine has tended to have the reputation in Britain of being largely restricted to 'yuppies' in the city, rock musicians and other denizens of the night-club scene, and exotic bohemian types. Amphetamine has always been far cheaper and more readily accessible for most people looking for a stimulant high. On the other hand, cocaine has always had its place in the hard drug sub-culture. A recent study showed that the number of heroin addicts using cocaine increased from 13–29 per cent between 1987 and 1989. In the same period, those taking their cocaine in the form of crack went up from 15 per cent to 75 per cent. The increase in numbers of cocaine addicts notified to the Home Office (1085 new addicts in 1990) has not kept pace with the rising seizure rate, suggesting that the large majority of cocaine users are 'invisible'; either they are not experiencing problems, or they are not presenting the problems they do have to doctors (who are the only professionals required by law to notify).

Recent estimates are that at least 30 million Americans have snorted cocaine at some time in their lives, with almost 50 per cent of adults between the ages of 25 and 30 having tried it at least once. There are six times as many cocaine addicts presenting to medical centres for treatment as heroin addicts. The cocaine epidemic was approaching its peak in 1979, at which time 10 per cent of those aged 18 to 25 randomly selected in surveys had used cocaine within the last 30 days. By 1986 it was estimated that three million Americans were using it regularly, with a total population prevalence of 3 per cent. It appears that, in the course of 1990, 6.2 million people in the US tried cocaine at least once, half of these went on to use it regularly, and at least 200 000 ingested it daily. The proportion of cocaine consumed in the form of crack in the US and Canada has increased very sharply in recent years. For example, two-thirds of the cocaine consumed by the youth of Toronto in 1989 was taken in this form. There may be as many as half a million regular crack smokers now.

Cocaine hydrochloride (referred to as coke, snow, charlie, Bolivian Marching Powder(!)) is usually sold as an off-white crystalline powder currently priced at between £50 and £80 a gram ($80–120 in US) depending upon the size of the deal and the buyer's closeness to the wholesaler. It has a bitter, tongue-numbing taste. Purity is variable, but averages between 50 and 70 per cent. Contaminants include related substances such as procaine, active but cheaper (and often more toxic) substitutes such as amphetamine or phencyclidine, or any suitable powders that come to hand as inert bulking agents.

Although it can be taken by mouth or, at the other extreme, dissolved in water and injected into a vein, most users arrange the powder into a 'line' on a hard surface and snort it into the nose through a tightly rolled piece of paper such as a bank-note, ideally of high denomination. Some people achieve the same result with a tiny silver spoon or other equipment specially designed for the purpose, amid the sort of ceremonial ritual that would seem excessive even to an unusually obsessional pipe smoker. An average line delivers a dose of around 25 mg cocaine.

As with all drugs, the effects are shaped to an important extent by the user's expectations, and the setting in which it is taken. In the case of street drugs, purity and the nature of contaminants are also important factors. Usually, a sense of well-being appears within a few minutes and grows rapidly. Confidence, optimism, energy, self esteem, and sex drive are all enhanced, and there is an overall feeling of exhilaration and happiness. Most normal pleasures are augmented but not distorted. The drug often enhances social skills, and part of its reinforcing property lies in the positive feedback which may be forthcoming from the user's companions. There is a reduced need for food, rest, and sleep.

Since cocaine is rapidly broken down in the body, the duration of action will tend to be measured in minutes rather than hours. Many people will content themselves with one or two lines, but some will seek to hang on to the high by snorting repeatedly. Once this pattern is established, decreasing sensitivity to the euphoric effects will necessitate rapidly increasing doses at ever briefer intervals. This may get out of control as an all-out binge, ending only when money or drugs run out, or exhaustion supervenes.

A survey has indicated that around 29 per cent of cocaine users are content to take the drug opportunistically if it happens to be around. A further 29 per cent buy their own supply but use it in a controlled, infrequent way. Twenty-eight per cent tend to use the drug more frequently and regularly, and find they devote a considerable amount of time to cocaine-related friends or activities. The remaining 14 per cent are likely to be compulsive, addicted users.

Freebase consists of cocaine stripped of its hydrochloride salt by a process of alkalinization and chemical extraction. *Crack* is a crude and impure form of freebase, easily prepared from cocaine hydrochloride and bicarbonate of soda in an illicit laboratory or somebody's kitchen, and so-called because of the popping and clicking given off by exploding impurities during smoking. It usually takes the form of soapy crystals known as rocks or chips, sometimes marketed in 'parcels' of four or more, and costing around £5 for a small chunk. It is smoked in a pipe or some form of improvised delivery system, such as a soft-drink can with holes punched in it.

The lungs are a highly efficient absorption system from which the blood carries its cargo directly to the brain without having to swish round the whole body or pass through the liver, where much of it would be broken down before it could have any effect. For this reason, crack produces sudden euphoria so powerful that words seem to fail people when they are asked to describe it. It delivers a 'rush' or 'high' comparable to that obtained by injecting the drug. Unfortunately, the physical and mental jolt associated with this is savage indeed. It is an extremely short-lived high lasting ten minutes or so, and is often followed quite rapidly by a very unpleasant down-swing of mood. Regular smokers feel irritable and 'wired', and an increasingly common method for dealing with this is to smoke or inject heroin. Many dealers in crack also sell opiates for this reason, and increases in crack use may therefore be sign-posted by an upsurge in heroin-related problems.

A number of people use cocaine in the context of multiple drug use, and they may well choose to inject it intravenously in order to maximize the intensity of the effect and the 'value for money'. It may be combined in the same syringe with heroin ('speedball') or a benzodiazepine such as temazepam to smooth off the rough edges of its effects by this route. Most people using a number of drugs intravenously, and many crack smokers, are leading a generally chaotic and dangerous lifestyle which will greatly augment the risks to their physical and mental health.

The most dangerous way to use cocaine is by smoking coca paste, a practice virtually unknown in the US and UK but becoming alarmingly prevalent in South America. Not surprisingly, injecting or smoking cocaine produces many more casualties than snorting it. When traffickers introduced freebase into the Bahamas in 1984, hospital attendances for problems associated with cocaine abuse increased seven-fold. Every year, smugglers (known as mules) stuff condoms or clingfilm packages with cocaine and swallow them. Should these burst or leak, a fatal outcome is by no means unusual.

It is estimated that 15 out of every 1000 American users of cocaine were seen in the emergency room of a hospital for drug-related problems in 1989. The yearly death-rate was reckoned to be one in every 2000 users (the figure for cigarette smokers was 12.6 per 2000 users).

Excessive doses are likely to produce sweating, dizziness, high body temperature, dry mouth, trembling hands, and a ringing in the ears. Anxiety and irritability may be evident, as may repetitive skin-picking and involuntary grinding of the teeth. The blood pressure may go up to the point of producing a stroke (bleeding into the brain which often results in paralysis). A direct toxic effect on the heart may cause it to beat irregularly and be less efficient, or stop altogether. Fits can occur which may lead to unconsciousness. Treatment of these emergencies is symptomatic as there is no direct means of neutralizing them. Injectors expose themselves to life-threatening infections, and many other hazards.

It is possible that long-term use may irreversibly damage particular dopamine-bearing nerves, or small blood vessels in the brain. More prosaically, the narrowing of the blood vessels in the noses of snorters can lead to a chronic runny nose, ulceration, or collapse of the nasal cartilage with striking effects upon facial architecture. Crack smoking is very hard on the lungs, and it is associated with severe chest pain of uncertain origin, asthma, and bronchitis.

There are reports of impaired hormonal and reproductive function in some long-term users, but of more concern are the effects of cocaine in pregnancy. This is seen as a tremendous problem in some North American cities where, in certain neighbourhoods, as many as a quarter of newborn babies test positive for the drug. Cocaine can interfere with the supply of oxygen and nutrients to the fetus through its constricting action upon uterine and placental blood vessels, and this effect may also result in premature labour or stillbirth. The drug has been shown to cause fetal damage in animal studies, but it is unclear whether this also occurs in humans. Some pregnant women who are using crack have been reported to be seriously undernourished or physically ill, or leading the sort of lifestyles which pose tremendous risks for themselves and the unborn child.

The phenomenon of 'crack babies' has received a lot of publicity. The offspring of crack-addicted mothers may be small-for-dates, irritable, trembly, poor feeders, and unresponsive to cuddling. These effects usually wear off in the course of a few days. The possibility of longer-term behavioural problems has been raised, but there are so many confounding variables (for example, social deprivation, poor nutrition and general health) that this is likely to remain uncertain. What is clear is the increased levels of domestic violence, physical, and sexual abuse in crack-using households.

Impulsivity, disinhibition, and impaired judgement may lead to disaster. High doses can cause anxiety, panic, irritability, or confusion. Hallucinations or delusions very similar to those seen in schizophrenia may occur. If the delusional beliefs take a persecutory form, there is a risk that dangerously aggressive behaviour may result without warning. Rarely, this loss of contact with reality (psychosis) persists even if no further stimulants are consumed.

Between 10 and 15 per cent of people who experiment with a snort of cocaine are destined to become compulsive users, usually within two to four years of the first exposure, but it is very difficult to predict with any confidence who is destined to fall victim in this way. Compulsive use of crack seems much more difficult to avoid, especially when taken in the context of social deprivation or emotional disturbance, but it is by no means the inevitable outcome as indicated by the tabloid press.

Compulsive use of both crack and powder often involves a series of binges lasting hours or days during which huge amounts may be consumed and ending only when supplies are exhausted or the user has a physical or mental collapse. At the end of such a run, three phases of withdrawal have been described. First comes the 'crash'. After a few hours of intense depression, agitation, and desire for more stimulants, the need for sleep becomes irresistable, and this may be induced with depressants such as alcohol, heroin, or benzodiazepines. Intermittent sleep may last as long as two days, possibly interrupted from time to time with avid guzzling of food. Next comes a period of low energy and motivation, depression, boredom, a lack of any sense of pleasure or enjoyment, which sometimes goes on for many weeks. This is likely to be accompanied by a powerful desire for cocaine which tends to come in waves and is brought on by contact with people, places, or things associated in the person's mind with previous drug-taking. If this desire is successfully resisted, the third phase consists of a gradual improvement in mood and the ability once again to find pleasure in ordinary things. The intensity and frequency of the bouts of craving steadily diminish.

Individuals in the grip of compulsive cocaine use—and

women crack smokers seem particularly vulnerable in this regard—may become involved in dangerous or humiliating sexual behaviour in the pursuit of supplies.

People with a history of psychiatric illness are more at risk of becoming compulsive users, and psychotic or depressive illness may be initiated or exacerbated by stimulants (and many other drugs). In the mid-1980s, one in every five people who committed suicide in New York was found to have cocaine in their bodies at the *post mortem*. An analysis of 300 psychiatric in-patients showed that 64 per cent could be categorized as 'substance misusers' of whom more than half were cocaine users.

Amphetamines

Amphetamine sulphate (speed) lacks the historical lineage and eloquent advocacy peculiar to cocaine. By and large, it is a rough and ready drug with a rough and ready clientele. Despite being regarded as a major problem in many countries throughout the world, it seems to attract relatively little coverage in the scientific literature.

Amphetamine was synthesized in 1887, but was not tested in humans until the 1920s. Amphetamine sulphate was first marketed as a nasal decongestant in 1932, and for use in asthma, obesity, pathological somnolence (narcolepsy), and depression soon afterwards. The unwanted effect of sleeplessness was recognized very quickly, but this did not inhibit the ever-widening range of suggested applications from Parkinson's disease, migraine, addictions, and seasickness to mania, schizophrenia, impotence, and apathy in old age (for which it seemed quite convincingly effective). This therapeutic enthusiasm was fuelled by many papers in the medical literature reporting very favourable results. In retrospect, the generally poor scientific quality of this work is all too evident.

The first non-medical application of amphetamine was to counter fatigue among soldiers in the Spanish Civil War, and in the Second World War this became commonplace in all armies.

Study Aids

In the British Army, it was to be used when the men '... were markedly fatigued physically or mentally and circumstances demand a particular effort', with a recommended upper dosage limit of 10 mg every twelve hours. Hitler was said to have been receiving regular injections of methylamphetamine (also barbiturates and other drugs) in the closing phases of the war, and this seems entirely consistent with descriptions of his emotional turmoil and erratic behaviour at the time.

The first recorded outbreak of widespread abuse occurred in Japan immediately after the war when large stocks intended for the military found their way on to the civilian market. By the fifties, pill misuse in the US was becoming commonplace, and the term 'speed freak' was coined. This did not inhibit the continuing expansion in the legal use of amphetamine, with weary politicians and tired housewives alike using it as a pick-me-up. It wasn't until 1956 that the first restrictions on use were introduced in Britain, but this did little to rein in demand. In the early 1960s, neurasthenia (chronic fatigue) was accounting for a quarter of all amphetamine prescriptions. In 1964, nearly four million prescriptions for amphetamine were issued in Britain, making up two per cent of all prescriptions written that year, and in 1971 twelve billion tablets were manufactured for medical use in the United States. In combination with short-acting barbiturates, amphetamines were found to be a useful aid to interrogation by operatives of various security services.

Pill misuse reached London early in 1960, and had captured the interest of journalists and other observers by 1962. A survey among Borstal boys at that time revealed that a third were regular users. A growing black market developed in amphetamine sulphate (Bennies), dexamphetamine (Dexies), methylphenidate (Rit), methylamphetamine (Meths, crystal), and Durophet (black bombers). A particular favourite among the Mods (a youth cult identified by a particular taste in music and dress-style) was Drinamyl (Purple Hearts), a potentially lethal combination of amphetamine and barbiturate. In 1964, the law in Britain was greatly tightened, but expansion of the black market sustained the growing epidemic. The prescribing of injectable methylamphetamine by doctors in the newly formed drug dependency units enjoyed a brief vogue, but several cases of psychosis among recipients caused this to go out of fashion very rapidly.

The risks were well-recognized by the streetwise. As one famous rock musician (Frank Zappa, quoted by Shapiro (1988)) put it: 'I would like to suggest that you don't use speed, and here is why: it will mess up your liver, your kidneys, rot out your mind: in general, this drug will make you just like your parents'. Restraints in the US during the seventies were associated with an apparent reduction in consumption, but this period also coincided with rapidly increasing availability of cocaine, a much more compelling reason for giving up amphetamine.

In Britain, the progressive withdrawal of legal supplies from the market encouraged the emergence of a succession of illicit local manufacturers. The volume of this production has consistently been sufficient to maintain amphetamine as the most widely used illegal drug in the UK after cannabis, although its quality and purity are generally very poor indeed.

Dexamphetamine sulphate is well absorbed when swallowed or sniffed into the nose. After taking it by mouth, peak levels in the blood are reached within an hour or two, and the level then halves every 12 hours or so. Its effect therefore lasts much longer than that of cocaine, and it is also less vulnerable to immediate neutralization on its first pass through the liver. Use in medicine is currently limited to the treatment of severely overactive children in whom it has a paradoxical calming effect, and narcolepsy (pathological sleepiness), with maximum oral daily doses of 40 mg and 60 mg respectively.

Methylphenidate (Ritalin) is now available in Britain only on a named patient basis and is virtually never prescribed, but is used more frequently in the US. It has a similar pattern of onset and peak activity, but is broken down in the body much more quickly.

As with cocaine, frequent ingestion causes rapidly diminishing sensitivity to the euphoric effects but not to the potentially dangerous effects on the heart, yet reported deaths due to overdose are rare. High blood pressure, and damage to small blood vessels in the eye, may occur in chronic users. It seems often to cause rashes, and heavy users may find their teeth rot because of a loss of dentine.

Amphetamine bears a close structural relationship to two of the brain's essential chemical messengers, noradrenaline and dopamine. The pharmacological effects are very similar indeed to those of cocaine, the only important practical difference being the much longer duration of action. If this is disguised experimentally, experienced users are surprised to find that they cannot tell the two drugs apart.

Animals will dose themselves enthusiastically with amphetamine in tests of dependency potential. This is inhibited by the dopamine-blocking drug pimozide, and long-term amphetamine use is associated with depletion of dopamine stores in the dopamine-powered 'reward pathway' referred to in the section on cocaine. Amphetamine is capable of producing irreversible damage to dopamine nerves in animals, but the significance of this for humans remains unclear.

Amphetamine stimulates the 'fight and flight' mechanism in the same way as cocaine, but the effect is more pronounced and prolonged. Its ability to induce powerful contraction of various sphincters led to a brief vogue in the treatment of bed-wetting children.

Potentially dangerous interactions occur with a variety of other drugs, including anaesthetics (heart irregularities), antidepressants (heart irregularities, soaring blood pressure), blood-pressure lowering tablets (antagonizes effect), and beta-blockers (soaring blood pressure).

Amphetamine taken in pregnancy may pose a risk to the healthy development of the baby's heart and bile system, has been linked with cleft palate, and is said to be associated with small-for-dates babies. Reliable information is hard to come by, however, and these effects are probably very rare.

In overdose, the elimination of amphetamine is much more rapid when the urine is acid. In hospital, this is achieved through the use of ammonium chloride, but the same result can be obtained when medical help is not at hand with large doses of vitamin C.

Amphetamine attracts little attention in the US media and

scientific literature at the moment, though in the past it has been calculated that more than 80 per cent of the world's illicit speed is consumed there. As the cocaine bonanza fades, perhaps amphetamine will emerge into the forefront once more.

In Britain, as mentioned above, amphetamine is second only to cannabis in the illicit drug market, having been firmly established as an endemic drug since the sixties. By the time they reach the age of 19 at least 10 per cent of the nation's youth has used it at least once. The number of seizures by the police has risen steadily from 2787 in 1987 to 6800 in 1991, and the quantities seized have increased spectacularly—from between 100 and 200 kg yearly in 1990, 1991, and 1992 to 550 kg in 1993. In 1990, 2330 people were charged with amphetamine-connected offences; by 1991 this had jumped to 3500. The average purity of seized amphetamine is incredibly low: 6 per cent in 1990, 10 per cent in 1991. The other 90 per cent consists of glucose, caffeine, milk powder, talc, or anything else that comes to hand of a convincing colour and texture.

Dexedrine (dexamphetamine) is marketed as a small white tablet, while illegally manufactured amphetamine sulphate (speed, whizz, sulphate) is usually sold in wraps or packages of whitish powder costing £10–15 per gram or less. A particularly pure batch was once dyed pink in order to identify it as a quality product, but odd colours are now much more likely to represent simply a marketing gimmick. The powder can be stirred into a drink, snorted into the nose, or dissolved in water for intravenous injection. Smoking it is less rewarding because the high temperature required to vapourize it destroys most of the active material. Smokeable methamphetamine ('ice') is available in some parts of Britain, but powerful side-effects are likely to limit its appeal. Nevertheless, if criminals eventually succeed in marketing 'ice' as effectively as they did crack, this will pose a particular threat to Britain with its large population of existing amphetamine users.

When inhaled into the nose, the powder often stings most unpleasantly, being likened on some occasions to snorting broken glass. Within a few minutes, there is an onset of effects which closely resemble those of cocaine: good cheer, vivacity, and optimism; increased energy and self-confidence; sharpened perception and concentration; reduced appetite for food and sleep. The main differences lie in the much longer duration of action, and the greater relative intensity of peripheral effects such as racing heart and dry mouth. Decreased sensitivity to the euphoric effects develops rapidly. The intensity of the effects will be limited by the purity of the material, which is likely to be very much lower than would be usual with cocaine. This difference may have given rise to the false impression that cocaine is a more powerful drug than amphetamine. When injected, a powerful rush or impact is noted within seconds, and both the mental and physical effects are much more powerful.

Irritability, suspiciousness, heightened aggression, and an unpleasant 'wired' sensation may be prominent, for the relief of which the user may turn to alcohol, sedatives, or opiates. Rapid changes in mood, flight of ideas, evident physical tension with restlessness are warning signs to companions or bystanders that impulsive violence may occur. As the drug wears off, the user is likely to feel depressed and washed-out.

The pattern of immediate and long-term unwanted effects is very similar to that described above for cocaine, except that the time between the escalating doses as the run builds momentum is longer because of the difference in break-down rate. Binges are followed by a short period of exhaustion and sleep, from which the user emerges into a phase of lethargy and inertia, often accompanied by anxiety or depression which may reach sufficient intensity as to induce thoughts of suicide or actual self-harm. The temptation to use more amphetamine is tremendous at this stage but, if successfully resisted, the mood can be expected gradually to return to normal. Occasionally, anxiety or depression persists for months or even years.

The original description of amphetamine psychosis by Connell in 1958 has become a classic paper in psychiatry. This alarming condition usually occurs in long-term high-dose users, but has been recorded after a single ingestion. It comes on a day or two after exposure, and consists of hallucinations (false perceptions which may be visual, auditory, or skin sensation, or any combination of these), disordered thinking, and delusions of persecution. These experiences seem completely real to the sufferer. Pointless, repetitive behaviours may occur, and there may be involuntary picking and scratching at the skin. These symptoms usually disappear gradually within a week of abstinence from amphetamine, but occasionally persist for much longer or become indistinguishable from schizophrenia.

Tolerance (diminishing sensitivity to the drug) causes many regular, long-term users to build their doses up to several grams of street material daily. Seesawing moods, poor concentration, insomnia, fluctuating suspiciousness, and a sense of being persecuted are quite common. Heavy consumption of alcohol, benzodiazepines, or opiates may represent an attempt to overcome these effects. A physical withdrawal syndrome is not prominent on cessation of amphetamine use, but depression, fatigue, lack of pleasure in life, extreme craving for drugs, and sleep disturbance are very common and may last for weeks or months.

Miscellaneous stimulants

Khat The leaves of the tree *Catha edulis* can be chewed or made into a tea. They contain a number of active ingredients, the most important of which is cathine. This compound has similar reinforcing properties in animal experiments to cocaine.

The leaves are cultivated over large areas of East Africa, where their use has been customary for centuries. Like coca leaves, they induce a sense of energy and well-being, but are also prone to cause stomach upsets, irritability, or sleeplessness. The risk of psychosis and compulsive use is well-recognized locally.

It is possible to buy khat in Western cities such as London and New York, but since the leaves must be fresh to impart their psychological magic, diarrhoea may be the only consequence noted by the consumer.

Pemoline (round, white, bi-convex tablets marked P9) is a milder alternative to amphetamine in the treatment of overactive children with a similar duration of action. In addition to the usual amphetamine-like unwanted effects, it sometimes proves damaging to the liver and bone marrow.

Some *appetite suppressants* act by exerting mild amphetamine-like effects, and thus have some value on the black market. They include diethylpropion, mazindol, and phentermine.

5
Psychedelics and hallucinogens

The defining characteristic of this group of drugs is their ability to induce profound changes in sensory perception, patterns of thinking and emotion without at the same time clouding the mind. Terminology is difficult, because it always seems to be judgement-laden; one person's *psychotomimetic* (psychosis-mimicker) is another's *psycholytic* (literally, mind-loosener, and by implication, consciousness expander). *Hallucinogen* is a widely used label which fails to do justice to the complexity of the human response to these substances. It is perhaps most appropriately applied to crude natural preparations, or to those drugs which disrupt mental functioning to the point of delirium (a mixture of delusions, hallucinations, confusion, disorientation, and memory disturbance). For those drugs which allow the user to remain oriented and fully conscious, the word coined in the 1950s by the psychiatrist Humphrey Osmond in the course of his correspondence with Aldous Huxley seems very suitable. '*Psychodelic*' was derived from the Greek words *psyche* (mind) and *delos* (visible). It was rapidly modified to the etymologically unsatisfactory, but possibly less psychiatry-orientated neologism, psychedelic.

Naturally occurring hallucination-inducers have been used for centuries in a prodigious variety of rituals throughout the Old and New Worlds, but the serendipitous discovery of an immensely potent synthetic psychedelic in mid-twentieth century had an important impact upon Western society. It provided the motive force for a counter-culture which developed with great rapidity, ushering in a period of artistic creativity and carrying at its heart a philosophy of life which posed a short-lived but formidable challenge to the existing world order. But how rapidly it all turned sour.

Hallucinogenic plants, along with fermented drinks and cannabis, have been used for thousands of years in social and religious rituals, for healing, and as a means of brief escape from a life which might well have been, as in medieval Britain, 'nasty, brutish and short'. Indeed, some people think that occasional periods of altered consciousness are a necessity for mental health (see Chapter 1), though there are of course many ways to achieve them without recourse to drugs. In modern life, such experiences have been associated with listening to music, meditating, having sex, and even jogging.

Solanaceous plants were the source of a number of these hallucinogens, often of peculiar toxicity. Datura, pituri, mandrake, and henbane, reeking as they do of the sorcerer's art, are rich in substances which interfere with the action of acetylcholine, one of the brain's important chemical messengers. The Europeans who first settled North America discovered such a weed which had been widely used by the native population for centuries. They christened it Jamestown Weed, and as jimsonweed it is still well known today. Such substances induce a delirious state, but may also produce loss of muscular control, temporary blindness, and paralysis. Many of their less toxic derivatives still find a use in medicine today.

More than a hundred psilocybin-containing mushrooms grow freely world-wide. The Aztecs were familiar with the easily recognizable, bright yellow Liberty Cap, which they called the 'flesh of the Gods' but they were also well aware that the safety margin of many other hallucinogenic mushrooms, for example *Amanita muscaria* or 'fly agaric', is dangerously small. These mushrooms produce a similar range of effects to LSD, although less intense and shorter in duration.

Harmaloid alkaloids, found in the seeds of various shrubs and vines growing in Arabia and Southern and Central America, provided the kick in a whole range of hallucinogenic snuffs and drinks still used ritually today. Ibogaine, containing similar compounds, remains central to the ceremonials of certain African societies. The search for new sources of such material left no stone, or creature, unturned; the fact that the skin secretions of certain toads were rich in a powerful hallucinogen did not escape notice. This produces a trance state awash with rich visual effects lasting for around six hours. Pleasure may be somewhat constrained by paroxysmal vomiting or vertiginous dizziness. Ibogaine, which was described by European missionaries in Africa in 1860 and extracted from root material in 1901, is yet more powerful and long-lasting, with enhanced toxicity to boot.

Highly toxic mescal beans, psilocybin mushrooms, lysergic acid diethylamide (LSD) containing seeds of the morning glory plant, and the mescaline-containing tips of the peyote cactus were used ritually in the Americas thousands of years before the birth of Christ. Partially suppressed for centuries, a formidable peyote cult resurfaced amongst Native Americans towards the end of the nineteenth century. At that time, peyote enjoyed a brief vogue as a constituent of patent medicines and tonics. As a result of this interest, mescaline was isolated in the 1890s and quickly established a place in Bohemian circles. It also found a niche in psychiatric practice, mainly in the hands of people who were observing and trying to understand psychotic illness (loss of contact with reality). The use of peyote in the religious practices of the Native American Church has grown steadily during the twentieth century, and is now the only legally sanctioned use of hallucinogens in the Western world. Four or five cactus tops will produce effects similar in intensity and duration to a modest dose of LSD, though possibly with a greater effect on heart-rate and blood pressure.

The key event which was to unleash the phenomenon of psychedelia upon the world occurred in 1943 when Albert Hofman, a research scientist at the Sandoz pharmaceutical company, accidentally ingested a microscopic amount of a chemical he had isolated five years previously while searching for a new heart stimulant. He felt very odd indeed, and decided to explore its effect further by taking what he thought was a miniscule dose, 250 micrograms, of the new substance which he had labelled lysergic acid diethylamide-25 (LSD). Bearing in mind that anything over 50 micrograms is enough to produce a hallucinatory effect, the impact of the ensuing 'trip' upon Hofman can readily be imagined. Although at times he feared for his sanity and, consistent with the thousands of first-hand accounts that have since been recorded, experienced a mélange of the awesome, the inspiring, and the grotesque, he was in no doubt as to the significance and potential benefits of his discovery.

Hofman's report of his experience unleashed a flood-tide of scientific articles, books, and conferences. The medical and psychiatric use of LSD expanded rapidly with well over 100 000 patients in the US and Europe receiving it during the fifties. Indications included depression, alcoholism, physical symptoms with no discoverable cause, tiredness, chronic pain, comfort for the dying, and, most frequently of all, as an adjunct to various forms of psychotherapy. North America was particularly open to this application since psychoanalysis was in its ascendancy at that time.

The CIA was most interested in other applications of a decidedly non-therapeutic nature. There was particular interest in an agent which might be able to facilitate interrogation or the reprogramming of those who had been brainwashed. Field tests included the administration of LSD to agents and others without their knowledge or consent; on at least one well-documented occasion, this had a fatal outcome when a man threw himself out of a window whilst in the grip of a prolonged psychotic state a couple of weeks after receiving the drug. In fact, some of the more mainstream scientific explorations of the time were rather dubious from an ethical viewpoint, using as they did such subjects as prisoners, autistic children, and psychiatric patients whose mental state made it impossible for them to understand properly the nature of the experiment.

Many hideous words were coined to describe these new drugs including *phantastica*, *psycholytics*, *psychotogenics*, *psychodysleptics*, and others with even more syllables, as well as Humphrey Osmond's *psychedelics*. Osmond was an imaginative psychiatrist working in Canada at the forefront of clinical research with LSD. Visiting California, he got into correspondence with Aldous Huxley who agreed to act as an experimental subject in an investigation and analysis of the subjective effects of mescaline. The resulting experiences were published in the book *Doors of Perception* (1954), the title taken from the quotation by William Blake: *'If the doors of perception were cleansed, everything will appear to man as it is, infinite'*. Huxley thought that mescaline '... lowers the efficiency of the brain as an instrument for focusing mind on the problems of life on the surface of our planet'. This interference with the brain's efficiency '... seems to permit entry into consciousness of certain classes of mental events, which are normally excluded, because they possess no survival value'. Huxley believed that the relentlessly materialistic focus of modern Western society had robbed its citizens of the spiritual dimension without which they could never rest content. He came to believe that psychedelics could bring this psychic contentment within the reach of 'everyman': that permanent and beneficial changes in attitudes and values would result from their use by ordinary people.

The use of psychedelics now spread rapidly beyond the little world of clinicians and lofty aesthetes into the grasp of earthier souls. Writers such as Ken Kesey and Allen Ginsberg became high-profile users and protagonists, and William Burroughs boasted his experiences with yagé. Jazz musicians such as Thelonius Monk, Dizzy Gillespie, and John Coltrane were into this scene very early, but classical musicians also experimented with the effects of psychedelics on musical creativity and interpretation. Timothy Leary, an unconventional Harvard psychologist, took psilocybin mushrooms in 1960 and was 'swept over the edge of a sensory Niagara into a maelstrom of transcendental visions and hallucinations'. Feeling that his life had been changed fundamentally and irrevocably, he began a series of psychedelic experiments with friends, colleagues, and students. His catchphrase, 'turn on, tune in, drop out', became a mantra for the hip generation and an excoriant for parents, teachers, and the rest of the so-called 'straights'. He also drew attention most effectively to the importance of set and setting in determining the effects of drugs—the attitudes, personality, and expectations of the taker, and the nature of the environment in which the drug is taken. He launched a personal attack on convention and ritual. Accused of being a psychopath, he riposted that 'there is no such thing as personal responsibility'. Introduced to LSD by Allen Ginsberg and Michael Hollingshead, he also gained access through them to the musicians, writers, and millionaires who were spearheading the new cultural movement. The scandal generated by the Harvard connection was important in gaining the interest of the mass media and nurturing the growth of the hippy movement.

Sacked from Harvard in 1963, Leary founded with others the International Foundation for Internal Freedom (IFIF) to pursue the study and propagation of psychedelia. He foresaw a time when every student would be 'turned on' in order to spearhead a new world order. Expelled from his base in Mexico, he returned to the US to set up the Millbrook Foundation in 1964 with the backing of a stockbroker friend. Meanwhile, Michael Hollingshead had set up the World Psychedelic Centre in London, patronized by the hip rock musicians and street poets of the emergent swinging London. The dissolution of the amphetamine-driven Mod scene in favour of the hash and acid atmosphere of hippiedom had begun.

On the proceeds of his novel *One flew over the cuckoo's nest* (1960), Ken Kesey bought a large estate close to San Francisco which became a focus and magnet for like-minded writers, musicians, and acolytes, and a setting for the notorious 'acid tests' in which gatherings of people drank Kool-Aid spiked with LSD and talked, played music, or made love. A folk-rock band called The Warlocks were inseparable from these happenings. Later they changed their name to The Grateful Dead and spearheaded the West Coast acid rock scene. Kesey's riotous and creative band of 'merry pranksters' voyaged in their ancient fluorescent bus in a 'stoned' odyssey across America, shocking and outraging bystanders and recording it all on 48 hours of uneditable film. They journeyed to Millbrook for a meeting of minds with the Leary group, but found them disappointingly staid and serious! Anyone interested in these extraordinary times should not miss Tom Wolfe's account in *The Electric Kool-Aid Acid Test* (1969).

The hippy philosophy had at its heart an unstable mixture of straightforward pleasure seeking, rejection of existing cultural values and conventional morality, snippets of Eastern religion, and repudiation of personal and institutionalized violence. Its epicentre in the early days was in the Haight-Ashbury area of San Francisco, but its cultural energy washed through England and mainland Europe. The psychedelic movement injected its unique style of music, art, and life-view into almost every level of Western culture. Scarcely a single segment of society could avoid reacting to it, even if this reaction took the form of disparagement or contempt. But even a politician's flowery multicoloured tie proclaimed it. It generated its own journalistic voice: *Berkely Barb*, *International Times*, *Oracle*, *Oz*.

The Grateful Dead, Cream, Jefferson Airplane, Pink Floyd, Soft Machine, The Incredible String Band, Traffic, Donovan—the list of musicians heralding the psychedelic influence is endless. The Beatles' musical style changed strikingly as they became caught up, as is clear from the album '*Rubber soul*' onwards. Perhaps the peak, the heyday, of the hippy era was represented by the 1967 'summer of love' in Northern California, where flower power found its finest hour.

The consumption of LSD skyrocketed within this cultural revolution. In 1962 it was estimated that 25 000 Americans had tried LSD, and by 1965 this figure had leapt to four million. In the same year, the first federal law controlling manufacture of LSD stopped short of outlawing individual possession, but more restrictive state laws soon followed and created the climate for the nationwide ban introduced in 1968. Britain had already banned it in 1966, and it was in this year that Sandoz ceased production for the licit market. These developments ushered in the era of the entrepreneurial underground chemists some of whom, such as Augustus Owsley Stanley III—'the man who did for LSD what Henry Ford did for the motor car'—were ideologically involved and not primarily driven by the profit motive. These individuals were often skilled enough to make a high-quality product, one of Stanley's more memorable vintages being immortalized by Jimi Hendrix's chart-buster, '*Purple haze*'.

After 1967, the tawdriness that was never far from the surface became more evident. The commercial exploitation of the hippy culture by both commerce and the criminal underworld grew rapidly in scale. Nowhere was this deterioration more clearly seen than in Haight-Ashbury itself, whose denizens had to cope with daily violence on the street over methamphetamine deals or poor quality acid cut with phencyclidine (PCP), while busloads of gawping tourists patrolled the area as if it were some kind of wildlife theme park. Ikons tumbled. In 1968, Timothy Leary was jailed on marijuana charges, and although his life remained no less colourful with escapes, recaptures, and eventual pardon, he never regained his former influence or prominence. Even the musical associations began to turn sour; at the Altamont Rock festival in 1969, the Hell's Angels stabbed and beat a man to death in front of the stage where the Rolling Stones were performing. This and the subsequent free-for-all were blamed on the mixture of amphetamine and bath-tub quality LSD freely available at the concert, although alcohol was quite possibly the chief culprit.

Scare stories surrounding LSD gathered pace in the media. Accounts of suicides, madness, blindness from staring into the sun, sexual orgies involving prominent people, and acid-head mothers giving birth to deformed babies began to change the public perception of psychedelics. Charles Manson shocked the world with his insane violence, his compulsive hold over his 'family' apparently greatly enhanced by their shared use of LSD. Enforcement of the law became more determined, and a number of underground chemists were caught and given long prison sentences in the early 1970s. From the middle of the decade until the early 1980s, LSD was in short supply and the trade in mushrooms and mescaline (usually PCP or similar artificial substitute in reality) took over. After this, the illicit industry recovered, and reasonably pure LSD has been available in most cities in Britain and the US to the present day.

Lysergic acid forms the nucleus of a group of chemicals called ergot alkaloids. Ergot is a fungus which, given half a chance, colonizes rye grasses, and has a fascinating history. When bread made from infected rye is eaten, two sets of symptoms may result. Most commonly, there is excruciating pain in the hands and feet due to constriction of the arteries which can cause actual death of the tissue (gangrene). The popular name for this affliction in medieval times, St Anthony's Fire, derived from the observation that sufferers who travelled to pray at the shrine of the saint often recovered. Cynics may surmise that the saint's influence must have been boosted by a lack of fungus in the region of the shrine. The other effect was to produce fits, hallucinations, coma, and not infrequently death through heart failure or depression of breathing. It was appreciated by the seventeenth century that ergot was the cause of these outbreaks, but epidemics have continued to occur well into the current century, the most recent being in France in 1953. Doctors found that some of the active ingredients in ergot were useful in the treatment of migraine and for speeding up childbirth, but there are now much less toxic alternatives.

Hofman synthesized LSD whilst searching among the derivatives of lysergic acid for a compound which might act as a respiratory stimulant. It was later found to occur naturally within the seeds of the morning glory plant. It induces peculiar behaviour in many different animals species, but is not self-administered by animals in tests of dependence or reinforcement. It does not seem to damage the brain or nerves and the lethal dose in animals is huge, death in such experiments being due to arrest of breathing.

LSD belongs to the group of substances called indolealkylamines, which are similar in structure to one of the brain's chemical messengers, 5-hydroxytryptamine (5HT, serotonin). The group also includes dimethyltryptamine (DMT), psilocin, psilocybin, the harmaloid alkaloids (such as William Burroughs' yagé), and ibogaine. The other major group of chemicals producing psychedelic effects are the phenylethylamines, which include mescaline, MDA, and MDMA (also known as Ecstasy). These last two have a distinct profile of activity and are discussed in Chapter 7.

Phencyclidine (PCP) was marketed in the 1950s as an anaesthetic free of respiratory or heart depression, but was soon restricted to veterinary use as its side-effects became apparent. The most notable of these was delirium and confusion on emergence from anaesthetic. It was completely withdrawn in 1965 as a response to an awareness of its widening abuse on the streets. PCP is a powerful painkiller, which both stimulates and depresses the brain. It is hallucinogenic but acts through different chemical pathways to LSD-like drugs and, unlike LSD, it is self-administered by animals in dependency tests. After an oral dose, it produces hallucinatory effects lasting about four hours, followed by a longer period of irritable depression, and remains detectable in the urine for several days. Ketamine (Ketalar) is a derivative of PCP which was synthesized in 1962 and is still in use today as an anaesthetic. It is shorter acting and safer than its parent, but unfortunately is now gaining a reputation on the street as an alternative to MDMA.

The psychedelic drugs produce changes in sensory perception, mood, intellectual and physical performance, and the pattern of thinking. It is difficult to have much confidence in the results of psychological testing carried out on those under the influence. Either the subject lacks the concentration or inclination to participate, or fluctuations in one modality will interfere with another being tested; for example, a sudden shift in mood from elation to despair would greatly influence performance in a problem-solving task. For this reason, reliable non-subjective information on the effects of the psychedelics on intellectual functioning is not available.

LSD is a white, tasteless powder which dissolves easily in water to a colourless solution. It is immensely potent in pharmacological terms, being hallucinogenic in doses from 50 micrograms (0.05 mg) upwards. When used as an adjunct to psychotherapy, doses between 200 and 500 micrograms were usual, and current street doses range between 50 and 250 micrograms.

Psychological effects begin to appear about 20 minutes after an oral dose, and the concentration in blood halves every five hours or so. The duration of action is variable but averages around 12 hours. Sensitivity to repeated doses rapidly decreases, but returns equally rapidly after a brief abstinence.

Only a tiny fraction of an oral dose actually reaches the brain, and even this has disappeared within 20 minutes. Since the duration of effect is many hours, the drug must cause some residual change in brain chemistry. This change seems mainly centred upon the serotonin (5HT) system, although noradrenaline (another vital chemical messenger) is also profoundly affected. As Huxley remarked, following the line proposed by the Cambridge philosopher C. D. Broad, the result seems to be an interference with the filtering and integrating function of the brain, so that the mind's eye is overwhelmed with a cornucopia of unfamiliar material. It also has a 'dehabituating' effect: the familiar regains its power to generate a fresh sense of novelty. It is interesting to note that sensory deprivation in a dark, soundproofed float tank greatly reduces the psychedelic potential of LSD, and totally blind people do not experience visual hallucinations or the so-called 'magic theatre.'

There is no physical dependence syndrome or withdrawal symptoms after ceasing regular use. Psychological dependence (see Chapter 10) is very rare.

The 'fight and flight' mechanism of the sympathetic nervous system is mildly stimulated. This may result in dilation of the pupils, modest increases in blood pressure and heart-rate, trembling, dizziness, loss of appetite, a small increase in body temperature, and sleeplessness. Death due to toxicity or overdose is almost unknown.

Despite the passing of the psychedelic era, such markers of prevalence as are available do not at all indicate a massive falling-off of use. In America in 1974, the proportion of college students who had ever used LSD was reported to be 23 per cent; the figure for a similar group in 1982 was 21 per cent (see footnote page 3). The number of new initiates to hallucinogens in America has remained remarkably steady over the last decade, averaging 613 000 new users each year. Seizure rates in Britain were low in the early 1980s, but later in the decade perked up sharply from 295 (84 000 doses) in 1986 to 1772 (295 000 doses) in 1990. This re-emergence coincided with the development of the 'acid house' music scene. Most of this LSD is apparently made on the west coast of America and then routed to the UK via Holland. It is estimated that one per cent of the overall British population, and 8 per cent of young adults, have taken LSD at least once. A small informal survey among state-school sixth formers in Oxford revealed that 10 per cent were current, albeit occasional, users.

PCP was very widely abused in America in the sixties, but went into a decline as its toxic potential was appreciated on the street. It has re-emerged since the seventies and is once again widely available. Although rarely described as a drug of first choice it remains a popular stand-by when other more desirable substances are in short supply, and finds its way covertly into many other street drugs as a cheap active bulking agent.

Fortunately, the use of PCP has never really caught on in Britain. At the time of writing, there are isolated reports that it has been identified as a contaminant of MDMA (Ecstasy) on sale in English nightclubs. More recently in Oxford, people who swear that their only recent drug use has been of cannabis have sometimes tested positive for PCP, suggesting that the American practice of boosting indifferent hash may be catching on here.

In years gone by, illicit LSD was often available as a clear liquid in foil-wrapped bottles to protect the precious elixir from the light, or absorbed into sugar cubes. There were also bits of impregnated paper (blotters), tiny tablets (microdots), gelatin sheets, and powder. An average dose was around 200 micrograms of LSD. Although often quite pure, it was sometimes contaminated with toxic by-products of manufacture, or more rarely adulterated with atropine or amphetamine. Currently, LSD is usually sold in Britain from sheets of impregnated paper squares of various sizes and adornment costing roughly £3 for a single dose. LSD content is variable but tends to be rather lower than in previous years, averaging 75 micrograms. There are periodic scares when it is announced in the press that LSD-impregnated stickers or transfers are being sold to children, presumably in the hope of getting them hooked. It is unclear whether this is of genuine concern or an elaborate hoax, but the drug's profile would not lead one to anticipate a successful outcome to this sort of criminal marketing strategy.

Possession or consumption of magic mushrooms (usually Liberty Cap) is not illegal in the UK, as long as no attempt is made to process the fungus in any way which extracts its active ingredient (psilocybin) or heightens its potency. Those who lack the energy or initiative to hunt them out personally can buy enough dried mushrooms (which *are* illegal) for a mildish trip (around 30) for a few pounds. They are occasionally sold ground up as a brown powder or pressed into tablets. Of course, one can never be at all sure what the powder or tablets contain. At the very least, there will be unease at the possibility that the picker may have selected unwisely or indiscriminately from the 3000 or so varieties of fungus which grow wild in the British Isles. Indeed a handful of these species are extremely poisonous. Occasionally, 'mescaline' is also available as a dark brown powder, but this is likely to be heavily adulterated if indeed it contains any genuine material at all. In America, it will probably turn out to consist largely of PCP. Users who try morning glory seeds obtained from garden centres in the UK risk toxic side-effects since they are routinely doused in fungicide.

The expectations and mood of the user and the setting in which any psychedelic drug is taken have a profound influence on the experience that results, as Timothy Leary pointed out. There are many hundreds of subjective accounts in the literature; a good selection can be found in Grinspoon and Bakalar's *Psychedelic drugs reconsidered* (1979) (see bibliography). Typically, things get underway about 20 minutes or so after an oral dose. After an initial sharpening of perception, visual distortion and a sense of detachment, apprehension, elation, or fear are the first signs of take-off. Thoughts begin to follow strange pathways, sparking off rapid shifts in mood. After this the shape of the experience becomes almost infinitely variable. Time seems to stand still.

Regular attitudes or viewpoints dissolve, and the familiar and hackneyed become novel, fascinating, or terrifying. Not only do hallucinations, illusions, and distortions of sight or sound tantalize or terrify, but one sense may seem to blot out all the others, or they may blend together so that a voice is felt or music experienced as a visual sensation. When the eyes are closed, swirling colours or images fill the mind (Timothy

Leary's 'magic theater'). Memories long suppressed spring into consciousness. Body image may be grossly distorted, or boundaries between the self and the external world seem to evaporate. Objects may seem to loom immense or shrink away, with loss of perspective; one user standing on Primrose Hill in London felt he could reach out and touch the Post Office Tower several miles away.

This psychedelic phase, with its twists and turns, its dread and delight, its mixture of pain and wonder, ebbs and flows in intensity. After a few hours, there is a gradual drift into a more relaxed, sensual phase. Every random thing or event has a meaning, a subtle significance. Gradually this magical aura recedes and small windows of ordinary reality obtrude and expand until all that is left is the memory of the journey.

PCP (also known as angel dust, hog) is usually smoked, although some people stir it into a drink, others snort it, and the extremely unwise inject it. Many Americans take it inadvertently when it is sold as mescaline or psilocybin, cut into other drugs as a filler, or sprayed on to indifferent cannabis to give it a bit more bite. In a recent screen of attenders at an American drug clinic, only 14 out of 100 people whose urine tested positive for PCP were aware that they had taken it.

Whichever way it is absorbed, PCP gives quite a jolt to the system. For the infrequent user, 10 mg is an active dose but regulars find they need bigger and bigger amounts to get the same effects. Within 15 minutes or so, a warm relaxed feeling gives way to dreamy euphoria, with or without hallucinations, lasting for several hours. Tingling in the limbs precedes a numb feeling and insensitivity to pain, whilst coordination and self-concern are greatly impaired. These effects gradually pass off to be replaced by a lingering sense of irritable depression.

Ketamine (Ketalar, 'Special K') is a white, smelly powder usually pressed into tablets for street sale. It produces effects which are similar to those of PCP but much shorter lasting, and is probably less toxic.

Most users of psychedelics allow weeks or months to elapse between trips, but some have runs of a few days at a time. Occasionally, one comes across people who have been in the habit of using high doses of LSD regularly for long periods, and they may seem to have only a rather tentative grasp on reality. However, they are generally easy-going and affable, and such investigations as were done in the sixties and seventies suggest that their intellect and brain functioning remain pretty largely unimpaired. The sort of compulsive use that may be associated with stimulants or opiates is almost unknown. These are not addictive drugs, it just isn't that sort of experience.

The psychedelic drugs have varying risk profiles. Death from LSD overdose is incredibly rare, if indeed it ever occurs. People have survived vast inadvertent overdoses, and fatal cases usually seem to involve other drugs as well as LSD. Physical collapse following toxic doses also seems very uncommon. There is the occasional report of fits following a huge dose. Deaths resulting from accidents sustained under the influence of LSD are well documented, especially when the drug has been given to someone unawares. These may be related to lack of self-concern, impaired performance of tasks requiring concentration and coordination such as driving, the acting out of fantasies such as a feeling that one can fly, or simply a sense of omnipotence and invulnerability. Reports of people going blind as a result of staring into the sun while 'stoned' seem largely apocryphal. There is no convincing evidence that LSD damages chromosomes or the unborn baby.

The most distinct risk is that one will have a 'bad trip'. As we have seen, any encounter with LSD is likely to have some frightening or unpleasant moments. Sometimes, however, terror or distress dominate the whole experience. A bad trip is much more likely if the setting is unsuitable, if there are no sympathetic and trusted friends ('guides') at hand, or if the pre-existing mood was a negative one.

A strong conviction that the mind is breaking up, that one is losing control, that irrevocable madness is impending may lead to intense agitation and panic. Thoughts may flow toward suspiciousness and a sense of being persecuted, which then shapes subsequent hallucinations. The destruction of mental defence mechanisms can unleash profound feelings of hopelessness, emptiness, or futility. Occasionally, such feelings may find expression in self-harm or suicide, or more rarely outwardly directed violence and even murder. Religious or other delusions may lead a person to mutilate themselves in a ritualistic or symbolic way, for example by emasculation or eye-gouging, but again this is very rare. Most people recover from bad trips in a day or so, but occasionally the effects last rather longer than that. It has been estimated that between 1 and 1.5 per cent of regular LSD users will seek professional help at some time as a result of a bad trip.

Much more serious are the prolonged psychological disturbances which can follow the use of psychedelics. These consist either of prolonged mental illness with delusions and hallucinations (psychosis), or persistent anxiety or depression. Some idea of the prevalence of these problems can be obtained from studying the reports of large-scale studies carried out when use of LSD in psychiatry was legal. Figures are variable, but prolonged psychosis seems to have occurred in between one and ten patients per thousand exposed, and from 0.5 to 1 per thousand experimental subjects. Bearing in mind the large numbers of people who regularly take LSD, this is by no means a negligible risk. Hallucinogens are certainly more clearly associated with psychotic illness than any other recreational drug. Suicide was reported in from 0.3 to 0.7 per thousand patients. Major complications with LSD occurred in just over two per cent of psychiatric patients who received the drug.

When prolonged psychosis does occur, it usually resembles schizophrenia in its pattern of symptoms, though wide swings of mood are sometimes more prominent than would be usual in that condition. Most cases gradually remit in the course of a few days or a week with symptomatic treatment only, but some people remain ill for weeks or months. It remains an open question as to whether these drugs are capable of inducing a long-term psychosis in a person who was previously completely normal. The current consensus is that this happens very rarely indeed, if at all. In most reported cases of illness persisting more than a week or so after the last exposure to the drug, it is possible to demonstrate vulnerability factors in the shape of pre-existing psychological abnormality, or a family history of psychotic illness. The outlook for these chronic psychoses is not dissimilar to schizophrenia itself; that is, a fairly high likelihood of further attacks. One suggestion is that some people have a latent genetic vulnerability acting through a defect in a particular brain system. Further exposure to the drug would be most unwise for someone who has reacted in this way. Persistent anxiety and depression usually respond in time to standard treatments, and again the drug's relation to causation is uncertain.

'Flashbacks' are sudden recurrences of psychedelic experiences occurring when no drug has been taken. About a quarter of LSD users experience them at one time or another, and some find them quite pleasant. Most people are rather frightened and disturbed by the experience, and may think they are a sign of impending madness.

Flashbacks may last for seconds or hours, and usually come on in association with tiredness, stress, or the use of another drug such as cannabis. They may be sparked off by the sight, sound, or smell of something or somebody associated in the individual's mind with LSD use in the past. A number of complicated neurological explanations have been put forward, but none are entirely convincing. In most people, the experience seems to bear a close similarity to a panic attack. There is no connection at all with mental illness and flashbacks usually become less and less frequent, disappearing completely within a few weeks or months of abstinence from hallucinogens. Occasionally, they coalesce into an almost continuous experience lasting weeks, months, or even longer; the term *post-hallucinogenic perceptual disorder* has been used to label this phenomenon.

PCP is a much more toxic substance than LSD in overdose, and there are many reports of delirium, coma, brain damage, and death. Muscle stiffness, difficulty in walking and speaking, and constant drooling from the mouth may result from damage to specific parts of the brain by impurities arising during illicit manufacture. Memory may be permanently impaired.

The combination of a strong sense of invulnerability and

omnipotence, insensitivity to pain, impaired judgement, and reduced tolerance of frustration, is fraught with danger for the PCP user and those who may come into contact with him or her. Unpredictable sudden violence or self-harm are by no means uncommon. Various forms of psychotic reaction are reported, usually brief in duration but occasionally becoming longstanding.

New hallucinogens will no doubt continue to appear from time to time. The latest in Britain at the time of writing is gamma hydroxy butyrate (nicknamed GBH). Not at the moment illegal, it was originally produced as an anaesthetic but abandoned because of its side-effect profile. Its effects are described as a combination of LSD and Ecstasy. Little is known yet of its potential for harm when taken by 'ravers'.

6
Inhalants

It is clear from cave drawings in Australia and Mexico, and from the writings of Greek and Persian scholars, that humans (and some animals) have not missed the opportunity to inhale naturally occurring gases or vapours found to induce changes in consciousness. Mystics, including the Delphic Oracle, used such emissions, along with other artifices such as fumes from perfumes and burning spices, to induce visions and enhance contact with the spirit world.

Modern recreational use of inhalants has its roots in the nineteenth century. At this time, pain relief in surgery relied upon alcohol, opium, cannabis, or even, for some unfortunate patients, concussion or partial suffocation. Sir Humphry Davy, writing at the very beginning of the nineteenth century to deplore this state of affairs, was the first to propose the use of nitrous oxide gas as an anaesthetic and pain-killer. He also extolled in eloquent terms the experience of inhaling the gas for pleasure, describing how one could induce exciting visions without untoward effects upon sleep or appetite. He introduced this pastime to a number of his affluent friends, and their enthusiasm was sufficiently unbounded for them to toy with the idea of setting up a 'nitrous oxide tavern'.

In 1844, a landmark event in the history of anaesthesia took place with the painless removal of a molar tooth from the mouth of one Dr Wells, who had prepared himself for the ordeal by breathing nitrous oxide in the manner recommended by Sir Humphry. Ether was first tried in 1846 and chloroform the following year. These early efforts were rather hit and miss, not least because of the potential toxicity of the latter two compounds, and there was some strenuous opposition on religious grounds. Despite this, the practice of anaesthesia gained

ground quite rapidly. The royal seal of approval was given by Queen Victoria, who was delivered of her eighth child in 1853 with the aid of chloroform.

News of the recreational possibilities of these substances spread quite rapidly among the upper classes, and nitrous oxide or chloroform parties became quite the thing during the remainder of the century. Medical people were particularly active in these pursuits because of their ready access to the necessary pharmaceuticals. There will always be doctors who succumb to the temptations posed by proximity to powerful mind-alterers, with varying degrees of discretion. Abuse of newer anaesthetics such as trichloroethylene and halothane sporadically comes to light, sometimes through the tragic consequences of their very high potency. Nitrous oxide still makes an appearance at medical student parties from time to time, and has very occasionally been available to the general public at concerts of popular music, conveniently packaged in balloons.

The first description of the phenomenon of petrol-sniffing reached the newspapers in 1942, but it wasn't until the late 1950s that the recreational use of solvents first mounted a wider stage of public awareness, as the practice caught on as a social habit among teenagers throughout the US. (A solvent is a chemical whose task it is to keep a product in solution or inactive until it is released from its packaging, at which point it evaporates completely and frees the product to achieve its purpose.) This new wave of sniffers were typically young, poor, deprived people, often from ethnic minority groups, who could not afford alcohol or other mind-altering drugs.

At the same period, another form of inhalant was becoming increasingly popular in show business circles, and soon afterwards within the gay community on America's West Coast. This was the pungent amyl nitrite, a volatile liquid which had been introduced into medical practice as early as 1867 for the relief of pain in heart disease, a benefit brought about by its ability to dilate blood vessels. It underwent something of a resurgence for this indication in the sixties, when it was available over the counter in pharmacies. It came in small gauze-covered glass ampoules or 'vitrellas' (poppers) designed to be crushed between finger and thumb to release the vapour. Apart from the usual inhalant 'high', poppers soon got the reputation of enhancing the joy of sex.

Amyl nitrite was made a prescription-only medicine in 1969, and its place in the pleasure-market largely taken by butyl-, octyl-, and isobutyl nitrite. Virtually the only remaining legal source of amyl nitrite these days is within emergency kits for the treatment of industrial cyanide poisoning. Isobutyl nitrite was and still is widely available in small bottles as a room odorizer, though trade names such as *Rush*, *Locker Room*, *Bullet*, *Thrust*, and *Lightning Bolt* leave little doubt that the manufacturers have more than an inkling of the true appeal of their products. Between 1973 and 1978, more than 12 million such bottles were sold in American discotheques, pornography shops, and hardware stores.

Volatile substance abuse, called 'glue sniffing' rather misleadingly since it actually involves deep inhalation, was being reported throughout the British Isles by 1970. As indicated by the historical perspective above, it is clear that inhalant abuse has always been endemic on a small scale throughout the world whenever opportunity has arisen. Some occupations apart from medical practice have carried a special risk because of the ready availability and daily contact with volatile substances. These include shoe-making, carpentry, printing, carpet laying, dry cleaning, painting and decorating, and hair-styling. It was in the early seventies that British public interest was first aroused on any scale and, as has been the case with other illicit drugs and activities, this publicity undoubtedly contributed crucially to the rapid expansion of the practice. In the UK, the habit spread most rapidly among teenagers in the larger towns, especially those in Scotland, Northern Ireland, and northern England. By 1980, surveys indicated that as many as 15 per cent of adolescents between the ages of 14 and 17 years in both Britain and the US had used inhalants at some time.

Alongside these alarming reports came the awareness that this could be a risky pastime. More than 300 solvent-related deaths had been reported in the US by 1972, and in 1978 alone 1800 American adolescents required emergency medical treatment in connection with solvent abuse. Between 1971 and 1983, 282 deaths were recorded in Britain and by 1990 this figure had passed the thousand mark.

Legal restrictions against solvent misuse have been enacted in several states of the US, but not proved helpful. In Britain, politicians made the sensible decision not to travel this route on the grounds that it would not deter use, might increase the hazard by pushing the habit further underground, and encourage resort to more toxic substances. Since the early 1980s, solvents have been regarded as drugs in the eyes of the law where driving impairment is concerned.

In 1985, the Intoxicating Substances Supply Act was introduced in the UK, making it an offence to supply someone under the age of 18 years with a substance the supplier knows, or has reason to believe, will be used to 'achieve intoxication'. Successful prosecutions have been few and far between.

Some examples of everyday products containing sniffable solvents are glues and adhesives (toluene, benzene, xylene, acetone); cleaning fluids (trichloroethylene, tetrachloroethylene, carbon tetrachloride); aerosols (fluorocarbons, hydrocarbons); petrol (hydrocarbons); rubber solution (benzene, n-hexane, chloroform); typewriter correcting fluid (trichlorothethylene); paint (toluene); varnish and lacquer (trichloroethylene, toluene); nail polish and its remover (acetone, amylacetate); dyes (acetone, methylene chloride); fire extinguishers (fluorocarbons); room fresheners (butyl and isobutyl nitrites).

A typical solvent such as toluene is rapidly absorbed when the vapour is inhaled, and has a high affinity to the fatty tissues throughout the body. It enters the brain very quickly. Toluene disappears from the blood within six hours of the last inhalation, but may reappear days later as it is released from fat stores. A small proportion evaporates away through the lungs, but mostly the body has to get rid of it by breaking it down in the liver, then excreting it in the urine as hippuric acid. The highest acceptable concentration in the air for industrial workers is 100 parts per million; 'sniffers' can exceed this level many hundred-fold. Once the toluene level in blood reaches the modest level of 0.5 micrograms per gram, it can easily be smelt on the breath.

Recent research suggests that the general anaesthetics used in surgery knock you out through an effect on the lipid (fat) component of the membrane surrounding nerve cells, selectively altering its permeability to various chemicals. The balance of these chemicals across the membrane determines whether or not the nerve is receptive to transmitting an impulse. The solvents may achieve their euphoric and other effects by a similar mechanism.

Surveys carried out over the years suggest that the prevalence of sniffing by adolescents between the ages of 14 and 17 years has remained fairly constant. Most indicate that between five and ten per cent have inhaled a volatile substance at some time, and that approximately one per cent are current sniffers (see footnote page 3). It is evident that many millions of very young people throughout the world inhale solvents year upon year.

Although individuals from deprived backgrounds, ethnic minorities, and inner cities are over-represented in most samples, glue sniffing occurs among people from all social backgrounds. The prevalence seems to ebb and flow in different areas and at different times, a finding consistent with the social nature of the practice. There is some indication that it is more widespread in the summer months. Epidemics in closed societies such as boarding schools and prisons are reported from time to time.

The commonest age to sniff for the first time is between 12 and 14 years and users under eleven are rare, though the youngest ever reported was only four years old. As with most illicit drug use, almost everyone's first exposure is initiated and supplied by a friend rather than a stranger. Reports vary, but the practice seems to be somewhat commoner amongst boys than girls; certainly it is true that many more boys suffer fatal consequences or require medical treatment.

Getting on for three-quarters of those who try sniffing only do so once or twice, a further 20 per cent sniff for a few weeks or months, and only about 10 per cent persist long-term. Heavy users are likely to be abusing a wide range of other substances as well as solvents, and are much more likely to be suffering from psychological problems or a seriously disrupted home life.

The extent of use of the volatile nitrites is difficult to quantify, but it is clearly considerable. Millions of bottles of butyl nitrite are sold quite legally each year in Britain and the US.

The average home contains literally dozens of solvent-containing products at any one time, so the would-be sniffer has plenty of choice. Typically, a globule of suitable material such as glue or rubber solution is dropped into an empty crisp packet or small plastic bag which is then held over the mouth and nose so that the fumes can be deeply inhaled ('huffed') ten or more times. This is very frequently a group activity, and the bag is passed round from hand to hand. 'Sniffing' is not a term which accurately conveys the vigour of the technique.

Alternatively, a volatile liquid can be inhaled directly from the container, or poured on to a rag or coat-sleeve for easier or more surreptitious access. Sometimes the user may seek to enhance the effect by enclosing the head completely, for example under the bedclothes, or within a larger plastic bag. Suffocation or overdose are all too possible in such circumstances.

Small brown bottles of butyl nitrite or similar are easily available from some hardware stores, sex shops, and gay clubs in Britain and America costing a few pounds or dollars. If kept in the fridge with the cap firmly applied between episodes, a single bottle will provide several highs before completely evaporating.

Gas lighter fuel (butane) can be sprayed into a balloon or bag before inhalation, but all too often the plunger is pressed against the teeth and the contents projected straight down the throat. Aerosols, for example spray paints, are usually inhaled directly, but it is impossible to avoid absorbing droplets of paint (or whatever dissolved material the product contains) along with the solvent vapour.

What are the reasons for experimenting with inhalants? Curiosity, boredom, desire for a new sensation, the attraction of risk, pressure from friends, a desire to shock adults; all these play their part. Some children who are frightened or unhappy may be seeking an escape from their misery. Temptation is difficult to resist when substances are so cheap and easily available.

Inhalants have a limited place as 'party drugs'. When taken on the dance floor, they may provide a brief blast of energy but unpleasant dizziness or nausea is always a risk. There are very distinct dangers in mixing physical exertion with huffing solvents, as will be described below. Volatile nitrites may be used specifically for sexual enhancement by both hetero-and homosexuals.

Logic dictates that those who find their initial experience with sniffing unpleasant or frightening will avoid repetition, whilst those who enjoy it may persist for a while. The large majority soon find that other priorities re-assert themselves, and they simply leave it behind.

Inhaled chemicals are whisked in the bloodstream from lungs to brain in a matter of seconds, so euphoric effects come on extremely rapidly with a distinct 'rush'. The user feels exhilarated and excited, high, disinhibited, and powerful. This is often accompanied by blurred vision, slurred speech, a buzzing in the ears, and marked clumsiness. Interesting visual illusions or hallucinations sometimes occur, and can often be steered or harnessed in an enjoyable way. Inhibitions disappear, and during sex orgasm is, or at least seems, prolonged and intensified. As with all psychoactive drugs, the user's expectations and the nature of the environment shape the experience very considerably.

These peak effects usually last only for a few minutes or so, and are followed by a more relaxed sense of well-being for half an hour or more. Some seasoned users have remarked that the intensity of the pleasurable effects increases with subsequent exposure, contrary to experience with most euphoria-producing drugs for which the opposite is normally the case. Tolerance (increasing resistance to the euphoric effects leading to consumption of ever-larger doses) is likely to occur in regular sniffers.

During the high, most users feel a bit dizzy and lightheaded, and are aware of their heart pounding along rather quickly. About a third report an unpleasant headache, there may be buzzing in the ears, and a few feel distinctly nauseous. Some experience chest or abdominal pain. Irritant fumes may cause coughing bouts, sneezing, and streaming eyes. If the concentration of vapour is intense, there may be disorientation, drowsiness, and even unconsciousness at very high doses.

After the high has passed, there usually follows a hangover period characterized by lethargy, depression, irritability, restlessness and agitation, poor appetite, and altered sleeping pattern. The user may experience chills, and a variety of aches and pains. The mood swings unpredictably; memory and concentration can be impaired. Parents of regular sniffers sometimes notice that their child seems more secretive or suspicious, that hobbies and established interests are being abandoned, that patterns of friendship are altering, and that school performance is deteriorating. Regular users at times develop spots or ulcers round their mouths or noses, cracked lips, and seem to suffer from a perpetual cold, sore throat, or bronchitis. They may exude a characteristic odour, spill marks are sometimes evident on clothes (especially cuffs) or around their room, and smelly bottles, bags, or rags may be left lying about. Regular, heavy huffers, or first-timers who have completely misjudged the dose, occasionally require hospital admission if they become very confused, have fits, or lapse into complete unconsciousness. Recovery is usually rapid (within a few hours) and complete.

One study did demonstrate that secondary school children who sniff solvents perform less well than those who do not on tests of vocabulary, IQ, and impulsivity. However, when measures of background social disadvantage were taken into account, the difference disappeared. Also, the performance deficit did not correlate with the amount or frequency of sniffing. The authors concluded that there was no evidence of neuropsychological impairment in this particular sample of glue-sniffing adolescents.

In the very large majority of sniffers for whom the activity is a one-off or transient phase, no long-term harm ensues. People who do suffer harm are much more likely to be those using large amounts over prolonged periods, with other drug use and risky or deviant behaviour contributing to the picture. Unfortunately, there are the occasional tragic exceptions to this rule; approximately ten per cent of solvent-related deaths occur in children sniffing for the first time.

Solvents contribute to more deaths in people aged under 20 than any other drug. More than 80 per cent of solvent-related deaths occur in this age-group, with 60 per cent of these younger than 17-years-old. The youngest fatality on record was only ten. Deaths are much commoner in boys than girls, and peak at 15–16 years. At a time when the general death rate in adolescents is decreasing, there is a steady increase in mortality associated with solvents. In the 10–14 age group, they have increased from 3.7 per million in 1983 to over 10 per million in 1988; for the 15–19 group, the respective figures are 17.9 per million increasing to 28.9 per million. This represents 4 per cent of all deaths in this age group, and 8 per cent of those due to injury and poisoning. Most deaths occur when the sniffer is alone, often at home. Currently, approximately 150 youngsters die annually in the UK as a direct result of this activity.

The use of butane gas or aerosols is particularly risky, and in recent years these substances seem to account for between half and three-quarters of all deaths ascribed to solvent abuse. Lighter fuel sprayed directly into the mouth may cause the throat to swell massively and result in suffocation. Aerosols deliver many other toxic chemicals as droplets in the vapour, and have a particular propensity to cause the heart to beat irregularly or stop altogether.

Probably as many as half of the deaths associated with glue sniffing are caused indirectly. The combination of disorientation, clumsiness, a sense of invulnerability, and lack of self-concern is obviously a dangerous one. Any alcohol taken at the same time greatly increases the risk. Injury from accidents, suffocation with the plastic bag or as a result of throat damage, sudden fire or explosion when people mix sniffing with smoking cigarettes, or choking on vomit whilst semi-conscious or comatose form part of the sad catalogue.

Most of the other acute deaths result from irregular beating and decreased efficiency of the heart. Many inhalants exert a direct effect upon the heart's muscle and nervous mechanisms, and the heart is made much more sensitive to the effects of adrenaline released in response to exercise or emotion; a sudden scare, or abrupt physical exertion can provoke a potentially fatal disruption of the heart's rhythm. For this reason, it is essential not to threaten or over-react to intoxicated sniffers lest such a reaction is induced.

There is a small risk of long-term damage to the liver or kidneys in heavy, chronic sniffers. Toluene and benzene can depress the manufacture of blood cells in the bone marrow, leading to anaemia, sometimes irreversible. Benzene use is thought to be associated with increased risks of cancer and leukaemia. The brain is also vulnerable to structural damage in such people who may tremble, have slurred speech, become clumsy and lose their balance, or develop memory problems. There may be injury to the nerves of sensation in the limbs, giving rise to 'glove and stocking' anaesthesia. Occasionally the optic nerve (which relays messages from the retina of the eye to the brain) can be affected, leading to progressive loss of sight. Additives such as lead in petrol may contribute greatly to the toxic potential.

Inhalation of solvents by pregnant women would not be prudent, since these chemicals freely cross the placenta into the baby's circulation. The absence of any conclusive evidence of damage to the unborn baby in these circumstances may simply reflect a lack of appropriate research.

Heavy use of solvents is commoner in people with demonstrable psychological problems, and was found in one survey to be the major factor in 3.5 per cent of referrals of under-16-year-olds to one London psychiatric hospital. It seems likely that these problems pre-date the solvent use rather result from it; there is no evidence that solvent misuse directly causes mental illness. The sort of person who is seeking to blot out unpleasant emotions or thoughts is likely to be a solitary, daily user of larger and larger quantities, who is a regular attender at general practices or hospital outpatients with repeated social or medical crises.

Intermittent and even quite regular sniffers do not experience a withdrawal syndrome on stopping, apart from the hangover described above. Very heavy regular users, on the other hand, may experience a cluster of symptoms and signs similar to those suffered by people dependent on alcohol: trembling, sweating, and agitation, sometimes progressing in those who do not receive treatment to disorientation, hallucinations, delusions, and fits.

7
Ecstasy

Hallucinatory amphetamines produce, as the term suggests, a combination of amphetamine-like stimulation and mild sensory distortion, but their particular characteristic is to induce in a diverse range of user types a general feeling of empathy and goodwill towards all mankind. The term 'entactogen' has been coined in an attempt to convey the essence of this experience; not exactly the sort of catchy label which might capture the imagination of tabloid leader-writers or other social commentators.

More than a thousand variants on this chemical theme have been synthesized, but the most important of these in current circulation are methylenedioxyamphetamine (MDA), methylenedioxymethamphetamine (MDMA, Ecstasy), and methylenedioxyethamphetamine (MDEA, 'Eve'). At present, Ecstasy is regarded in Britain as something of an epidemic drug and those who write about it tend to do so in rather emotional terms. Indeed, the wild and speculative nature of the coverage which appears in both lay and professional media makes it difficult for the non-partisan observer to form any sort of balanced view.

MDA was synthesized in 1910, then lay dormant until 1939 when it underwent testing in animals. It was launched in 1941 as a treatment for Parkinson's Disease, without any great success. Gordon Alles, discoverer of amphetamine, wrote about his experiences with MDA in the 1950s, and perhaps it was this work which suggested some military potential to the organizers of the US chemical warfare research programme. MDA (the 'love drug') became available on the streets from the mid-1960s, and was subsequently brought under the control of the American Drug Abuse Prevention and Control Act (1970) and the British Misuse of Drugs Act (1971).

MDMA (Ecstasy) was synthesized in 1912 and patented by the pharmaceutical company E. Merck two years later. It too underwent toxicology studies courtesy of the US military authorities in the fifties, and was available, quite legally, for recreational use from around 1970. This non-medical use grew steadily, and seems to have had a particular focus in Texas and California. In the UK an amendment to the Misuse of Drugs Act was introduced in 1976 to outlaw all amphetamine-like compounds, but in America the drug continued to go from strength to strength. One Californian laboratory alone produced 10 000 doses monthly in 1976, 30 000 monthly in 1984, and 500 000 in 1985. This enormous expansion may have been related to the easing of the cocaine epidemic and some eulogizing publicity for MDMA in journals such as *Time* and *Newsweek*.

In 1976, reports were appearing about the successful use of MDMA as an adjunct to individual, marital, and group psychotherapy. Leo Zeff, a psychologist who had used LSD in therapy sessions ten years earlier, found that it helped people to communicate their feelings more effectively and tolerate criticism. It was also said to be helpful in the treatment of drug and alcohol abuse. A number of publications appeared in the psychiatric literature over the next few years, most of which were highly favourable in their conclusions about the drug's benefits.

At the same time, concern in official quarters about MDMA's abuse potential was growing, and the American Drug Enforcement Agency (DEA) began to investigate the possibility of bringing it under legal control in the early 1980s. Counter-arguments were put forward by psychiatrists and others and the debate looked set to run and run; but in 1985 the DEA abruptly terminated the discussion by invoking the Comprehensive Crime Control Act to bring MDMA under Schedule 1 (the toughest restriction available) control. Numerous appeals brought a temporary reprieve, but DEA lawyers were able to argue that the reports of clinical success were not of sufficient scientific rigour to be able to justify their conclusions. MDMA was restored to Schedule 1 control in 1988, where it still resides. Ironically, it seems that the publicity surrounding these protracted legal disputes must have been partly responsible for a spectacular increase in the non-medical consumption of the drug. The ban in 1985 also led to a flurry of 'designer drugs', chemical variants on the theme which did not fall within the scope of the law, one of which was MDEA. This loophole was plugged in 1986 by the Controlled Substance Analogue Enforcement Act, which proscribes drugs similar in structure or psychological effect to those already banned.

As a result of their legal status, almost nothing is known about the human pharmacology of MDA and MDMA. Only one relevant investigation in a single subject is on record, dating from 1976. This indicated that the effect of an oral dose became apparent within 30 minutes, peaked at around 60–90 minutes, and had disappeared in about four hours. The experience was described as '... an easily controlled altered state of consciousness with emotional and sensual overtones'. MDA is reported to be longer-acting and a 'rougher ride' than MDMA. It is supposedly less euphoriant and more amphetamine-like, and had a reputation on the street of being less well tolerated by women than men; for example, dormant pelvic infection might be induced to flare up.

Animal studies indicate that the most important actions of these drugs are mediated through effects upon one of the brain's chemical messengers (neurotransmitters), 5-hydroxytryptamine (5HT, serotonin). MDA and MDMA both cause a massive release of 5HT, and then seem to inhibit its synthesis so that the brain becomes temporarily depleted. 5HT is thought to play an important role in regulating mood, sleep, aggression, hunger and sexual activity. The primary, sought-after effects of MDMA seem to be related to the initial flood of 5HT through the brain, and the residual hangover to the subsequent depletion.

It has been established in animals that MDMA and MDA are specifically toxic to 5HT nerves, producing degeneration of the nerve terminal. Levels of 5HT and its break-down products, and chemical markers of new 5HT production, remain low for weeks or months. After several months, some evidence of regeneration is evident, but it seems likely that at least a part of the damage is permanent. The degree of damage is related to the size and frequency of the doses but animal species vary considerably in their susceptibility. For example, monkeys are much more vulnerable than mice; in the former, damage has been seen at doses as low as 2 mg per kg of body weight twice daily, a level that is worryingly close to doses typically taken recreationally by humans.

The implications of these findings for human users are uncertain for two reasons. First, it is difficult to extrapolate from animal work in the absence of even the most basic pharmacological information or confirmatory experiments in human volunteers. Only one relevant study is available at the time of writing. This preliminary investigation of 5HT function in volunteers who said they were heavy MDMA users did demonstrate an apparent abnormality in their 5HT systems in comparison with non-MDMA using control subjects, but the difference was not statistically significant and requires confirmation. Another difficulty is that subjects such as these may well have used a variety of other drugs, and unrecognized contaminants within illicit MDMA may themselves be highly toxic.

Secondly, the functional or practical significance of this nerve damage is unclear. In other neurotransmitter systems, there is evidence of 'neuronal redundancy'—that is to say they have been endowed with many more nerves and brain cells than are actually needed to function adequately, in which case the loss of a few thousand may be neither here nor there.

Bearing these reservations in mind, there is more than enough evidence of toxicity in animals to underline the urgent need for research to clarify the effect, if any, on the 5HT system in humans. Most people would be keen to hang on to as many brain cells as possible, given the choice. However, it is important to be aware that inconsistencies abound in our response to evidence of drug toxicity. Fenfluramine, a legal drug used quite widely in both Britain and America as a slimming pill, is at least as toxic to nerve cells in the same brain system as may be affected by MDMA. This seems to arouse no concern whatsoever to the regulatory authorities.

MDMA gives positive results in animal models of dependency,

but there is no evidence of physical dependence in humans, and compulsive use or addiction seems very rare.

The amphetamine-like effects cause stimulation of the sympathetic nervous system, the so-called 'fight or flight' mechanism. Heart-rate and blood pressure go up, blood is diverted to the muscles away from the guts and other maintaining mechanisms, the body's chemical processes speed up, and more oxygen is absorbed through the lungs.

Certain pre-existing diseases, of which a person may not be aware, increase the likelihood of unpleasant or dangerous reactions to MDMA. People with any of the following would be well advised to stear clear of it: diabetes, liver disease, high blood pressure or heart disease, epilepsy, glaucoma, hyperthyroidism, or any form of mental disorder. Prescribed medicines may interact with it. Pregnant women would be wise to avoid MDMA: although there is no direct evidence of toxicity to the unborn baby, this may simply reflect the lack of appropriate research and there is always the risk of toxic adulterants in black-market drugs.

There are anecdotal reports that MDMA inhibits the immune system in some way, so that users become more susceptible to colds and other infections. This must remain speculative pending scientific confirmation.

The legal restrictions in both the US and Britain prevent doctors from prescribing it in any circumstances. Formal surveys of illicit MDMA use are few and far between. It is believed that 85 per cent of the MDMA now entering Britain is routed via The Netherlands. Seizure rates have increased from 399 in 1990 to 1735 in 1991. Perhaps more significantly, the quantity seized has risen from a few kilograms in 1990, to nearly 300 kg in 1991, and 550 kg in 1993. In London alone 66 000 tablets were confiscated in 1991, compared to 5500 in 1990. In a sample of London youth club members, 11 per cent had used MDMA, but informal surveys in some schools have indicated a much higher prevalence amongst 15–17 year-olds (see footnote page 3). At the time of writing, it is estimated that more than 500 000 people in Britain have used the drug at least once.

I am not aware of any very recent surveys from the US, but a

1987 survey at Stanford University suggested that 39 per cent of students had used MDMA at least once. Of these, about half had used it less than six times, a third between six and ten times, and 12 per cent more than ten times.

Black market MDMA is available as a powder, tablets, and capsules of various shapes, colours, and sizes, with rather whimsical names including Dennis the Menace, Disco Biscuits, Big Brown Ones, Burgers, M25s, Pink Scuds, Bug-eyed Billies, California Sunrise. Quality control is variable, with MDA and amphetamine turning up in confiscated samples and anecdotal reports of adulteration with many other substances. Occasionally there is no MDMA present at all, but the usual content is between 75 and 200 mg. At some nightclubs in Amsterdam it is possible to break off a tiny piece of tablet and have it submitted to an on-the-spot purity screen, but the scene in Britain has not yet reached this pitch of sophistication. The drug is almost always swallowed, but may occasionally be snorted into the nose or, very rarely, injected into a vein.

There are two distinct classes of user. By far the most numerous are young, energetic nightclubbers and 'ravers' who take it for the pleasure of the high and the enhanced sense of togetherness. Some of these will also be amenable to other drugs such as amphetamine, LSD, or cannabis, which are likely to modify the profile of effects. The other much smaller group is made up of more spiritually inclined people, perhaps exploring a 'New Age' philosophy of life, or seeking personal insights alone or within a relationship. It is unusual for anybody to take the drug compulsively, or even very frequently. It is bought for a particular time and place. If someone buys a tablet or two of Ecstasy in anticipation of a party or encounter session and the event is then cancelled, the drug is likely to be stowed away in a drawer until another suitable setting can be arranged. One should note, however, that some people do go to an awful lot of parties!

As with all drugs, the effects of MDMA depend to a large extent upon the characteristics of the taker, and the setting in which it is taken. It is hardly surprising that the experiences of a middle-aged, middle-class psychiatrist who takes MDMA in his living room with professional colleagues and a string quarter on the CD-player might contrast starkly with those of an unemployed adolescent raving the night away in a disused aircraft hanger with a thousand others, dancing to a sound system producing more decibels than a jumbo jet.

There are some common strands, however. Half an hour after taking a tablet, both the psychiatrist and the raver would have noted an increased sense of alertness, possibly accompanied by a dry mouth, skin tingling, and an awareness of the heart beating faster. A sense of closeness and empathy with other people gradually appears. A change in the quality of light perception (luminescence) usually occurs, and some people may experience actual visual hallucinations. The jaw often feels tight, sometimes with involuntary grinding of the teeth (bruxism).

Reports differ on the effects upon sexual behaviour. Concern has been expressed that MDMA and variants might release inhibitions and lead to an upsurge in promiscuous, 'unsafe' sex, alarming to moralists and those concerned about the spread of HIV. The drug certainly seems to enhance the sensual enjoyment of sex, but whether or not it increases sex drive remains unclear. The amphetamine-like component of the drug's activity may cause erectile failure in men, and inhibit orgasm in both sexes.

Some American psychiatrists have reported their personal experiences with Ecstasy. More than half described enhanced powers of communication, feelings of intimacy, improved interpersonal skills, and euphoria. Most also experienced enhancement of the senses (for example, increased musical appreciation), reduced fear and sense of alienation, increased awareness of emotions, and suppression of aggressive impulses. Some said that the experience had caused them to rethink their priorities in life, whilst others felt that their social functioning had been significantly improved. It is worth noting that a quarter of these people combined the use of MDMA with cannabis or tranquillizers.

In a sample of recreational users in Australia, the majority had taken MDMA several times rather than once only. More than half also smoked cannabis, and a significant minority were also on nodding terms with amphetamine, LSD, and amyl nitrite ('poppers'). A quarter had used cocaine at some time, but tranquillizers or opiates were restricted to a tiny minority. Alcohol consumption was modest and, consistent with other reports, seemed to interfere with the enjoyment of MDMA. This sample seems to have contained a predominance of socially outgoing types, possibly with a higher proportion of homosexuals than might be expected by chance, who took the drug out of curiosity in the anticipation of fun and with considerable peer approval. They were aware of the reported risks to physical and mental health, and were concerned about possible impurities and the lack of research on the compound. The vast majority took the drug orally, experiencing an effect usually lasting around five hours or so and a hangover which extended over the next 24 hours or more.

There is consistency in the reports from this and other groups that the most rewarding effects come from the first exposure, with diminishing returns for most people thereafter. Increasing the dose ('stacking') may enhance the rush and increase the likelihood of hallucinations, but more often it just increases the ratio of unwanted to wanted effects. These individuals reported positive experiences from the drug on almost every occasion they took it, but they also seemed to accept that some unpleasant side-effects were almost inevitable.

Those who had also experienced amphetamine and LSD had no difficulty in distinguishing the effects of all three drugs one from the other. Ecstasy was most clearly associated with intimacy, easy-going acceptance of others, and sensual euphoria. Amphetamine was seen primarily as alerting and energizing, with enhanced confidence and self-esteem, while for LSD the emphasis was on enlightenment, insight, and open-mindedness.

Anecdotal reports from rave-goers and night-club denizens mirror these reports of empathy, energy, and a sense of inner peace. Many of the small-scale dealers at street level share the ideological values of their customers, and the sense of an in-group looking out with weary contempt at the bewildered 'squares'—teachers, parents, professionals—is reminiscent of the earlier hippy generation. The importers and sellers higher up the pyramid take a more robustly commercial view of the situation. As with other illicit drug markets, the size of the profits precludes any of the finer feelings: a single seizure recently in Britain had a street value of well over one million pounds. This more than compensates for the larger penalties associated with Class A drug dealing (see Chapter 12).

As far as risk is concerned, one of the greatest is that you may not be taking MDMA at all, or that there is contamination with cheap but toxic hallucinogenic substitutes such as phencyclidine (see chapter 5) or damaging by-products of incompetent manufacture.

MDMA itself is associated with a high prevalence of minor unwanted effects including clumsiness, lack of coordination, and poor concentration which may lead to accidents; drowsiness or restless agitation; fear or anxiety sometimes escalating into a panic attacks or a sense of loss of control; depression; a racing heart, dizziness or fainting; nausea and vomiting; headaches; persecutory feelings or unpleasant hallucinations.

Most people experience hangover effects which last for a day or two. These usually consist of tiredness, lethargy, irritability, depression or seesawing emotions. The muscles may ache and feel stiff, especially around the jaw after all that tooth-grinding. For many people with jobs, these residual effects limit the use of Ecstasy to Friday or Saturday nights. Flashbacks have been reported.

There are well-documented records of serious physical and psychological reactions to MDMA. Whilst many of these seem to have occurred in people predisposed in some way or taking large doses of other drugs as well as MDMA, it is clear that idiosyncratic reactions to modest doses do occur, though with extreme rarity, and that these may prove fatal. Generally, such limited information as does exist indicates that serious side-effects correlate poorly with blood-levels of the drug. The amphetamine-like effect may overstretch the heart and circulatory system, especially if there is unrecognized narrowing of the arteries or other pre-existing disease, so the heart beats inefficiently or stops completely. Fits have been reported, and blood pressure changes may cause the person to have a stroke (cardio-vascular accident). Interactions with other illicit or prescribed drugs (for example, antidepressants of a particular type) can prove fatal. An association with liver damage has been suggested.

A rare but particularly serious physical reaction is a vicious rise in body temperature associated with damage to muscle fibres and blood-clotting inside arteries and veins. This can lead to liver or kidney failure, coma and convulsions, and death. There is a convincing theory that the environmental conditions to be found in nightclubs or at raves may increase the risk of this reaction. These places may be very hot, and with energetic dancing and a lack of water or soft drinks dehydration can easily occur. External heat, vigorous physical activity, and dehydration may combine with the effects of the drug on the body's temperature-controlling mechanisms to bring about a rapid increase in temperature to the point at which internal tissue damage occurs. It is therefore essential that at each venue there should be adequate ventilation with access to water and soft drinks, rest areas, and cool retreats where people can 'chill out'.

There is also the risk of injury or death resulting from the clumsiness and impaired judgement referred to above. It is clear that reactions to the drug severe enough to result in medical intervention are increasing; enquiries to the Guy's Hospital Poisons Information Service in London are reported to have gone up considerably in 1991.

Although these risks are undoubtedly genuine, it is important to keep them in perspective. Deaths and serious reactions are very rare indeed in relation to the huge amounts of the drug which have been consumed over the years. Somewhere in the region of 20 to 25 deaths world-wide have been attributed directly to MDMA at the time of writing but at least some of these, and other serious but non-fatal reactions, may be due at least in part to pre-existing physical disease, other drugs, or adulterants in the black-market supplies.

Serious psychological reactions to MDMA can also occur. These usually take the form of prolonged anxiety or depression, but a few cases of persistent psychotic illness with hallucinations, delusions, and loss of contact with reality have been reported. In almost all of these cases, an important contribution from pre-existing mental disorder, or the regular use of other street drugs clearly associated with the risk of psychosis such as amphetamine, cannot be ruled out. Again, these reactions seem few and far between when viewed in context of the millions of doses consumed worldwide. Treatment is symptomatic with anti-psychotic medication and psychological support. No outcome figures are available for these patients, many of whom will later be diagnosed as schizophrenic.

8
Tranquillizers and sleeping pills

Barbiturates

Insomnia is a difficult phenomenon to tie down because individual variations in the requirement for sleep vary so widely, but the need for that 'balm of hurt minds, great nature's second course, chief nourisher at life's feast' has preoccupied mankind from hunter–gatherer days. In modern Britain, up to 15 per cent of men and 25 per cent of women attending a GP's surgery will be complaining of inadequate sleep. Herbal potions, opium, and alcohol have all served as sleeping-draughts in their time. In the middle of the nineteenth century, the sleep-inducing property of bromide was discovered, and its derivatives became very popular. It was not until the 1930s that recognition of their toxic effects on nerves led to their disappearance.

The coming-of-age of organic chemistry permitted many pharmaceutical innovations before the turn of the century, many of which are still in use today. Examples from the sedative field include paraldehyde and chloral but the most significant in this domain was barbituric acid, and the subsequent marketing of barbitone in 1903. Phenobarbitone quickly followed, and quite soon there were more than 50 'barbiturates' on the market. These were prescribed more and more widely for insomnia and anxiety and later, in combination with amphetamine, for depression. Cases of dependence were reported in Europe as early as 1912, but it was not until 1950 that the possibility of becoming physically dependent upon barbiturates was fully acknowledged.

For many years, the morbidity and mortality associated with barbiturates was mainly confined to the large numbers of people receiving legitimate prescriptions from doctors, and those

members of their households who had access to the bathroom medicine cupboard. In the late 1950s, young people who had discovered the delights of amphetamine also discovered that barbiturates were just the thing for bringing the pace of life back down to a controllable tempo after partying with speed. Heavier users of stimulants came to realize that combining them with barbiturates mellowed the sharp edges of the high and eased the crash which eventually had to be faced at the end of a binge. Heroin users found that the ready availability and low price of barbiturates made them an invaluable fall-back when supplies ran short, and dealers cut them into street opiates as a cheap alternative to the real thing. Oblivion seekers found they had access to the ultimate lift-shaft.

This rapid growth in the non-medical use of barbiturates by young people was facilitated by the huge amounts then being manufactured for the legal market. In 1960, one billion tablets were dispensed from pharmacies in Britain. Drug combinations containing barbiturates found a special niche within the street scene. Purple Hearts (Drinamyl), a combination with amphetamine, were all the rage with hipsters and party-goers. Mandrax (methaqualone—not a barbiturate but very similar in profile, plus diphenhydramine) arrived in Britain in 1965, and by 1968 was the most widely-prescribed sleeping tablet. Although withdrawn in the 1980s, methaqualone retains a select but enthusiastic band of black-market devotees to the present day.

By the late 1960s, there were more than 2000 barbiturate-related deaths each year in Britain. The deaths of several well-known pop musicians including Brian Jones, Jimi Hendrix, Janis Joplin, even Elvis Presley were associated with taking barbiturates. At least a third of the heroin addicts attending a major drug-dependency unit in the US were also regularly injecting barbiturates, and more than half of London's opiate addicts had injected the stuff within the previous year. On the broader stage, these drugs were second only to coal gas as the instrument of death in suicide, accounting for 8000 deaths yearly in Britain.

These sombre statistics did little to stem the enthusiasm of doctors or their insomniac and anxious patients. British doctors wrote 16 million prescriptions for barbiturates in 1964. Five hundred thousand people were reckoned to be taking them legitimately, with almost a quarter of these dependent upon them. It was only the appearance and rapid expansion in use of a safer alternative, the benzodiazepines, that stemmed the tide. The barbiturates went into a rapid decline during the seventies, and their use in medicine became highly restricted. There was a small upsurge in illicit use in the early eighties, and they retain a limited though toxic presence within today's street drug scene.

The barbiturates can be divided into three pharmacological categories based on their duration of action. Short-acters such as thiopentone are never prescribed to patients outside the operating theatre or intensive care unit, are of no interest to recreational users, and will not be considered further. Long-acting drugs, which include phenobarbitone and allobarbitone, hang around in the bloodstream for many days and so rapidly accumulate in the body if taken regularly. Blood levels of the medium-acting group (consisting of pentobarbitone, quinal-barbitone, butobarbitone, and amylobarbitone among others) halve every 30–40 hours. All are absorbed well by mouth and are gradually broken down by the liver. This drug-neutralizing activity of the liver becomes generally hyperactive through its attention to the barbiturates, and this can result in reduced effectiveness of certain other drugs as the liver disposes of them more quickly. Examples of these include the contraceptive pill, blood-thinning agents, and steroids.

Resistance develops to the sleep-inducing and tranquillizing effect but less so to the depressant effect on breathing, which may lead to accidental overdose with fatal consequences. This resistance generalizes to other brain depressants such as alcohol.

Barbiturates facilitate the activity of the main inhibitory chemical messenger (neurotransmitter) within the brain, gamma amino butyric acid (GABA). Messages pass along nerves in the form of waves of electrical depolarization. There is an electric charge across the resting cell membrane of the nerve because of differences in concentration of various chemicals across it, and these differences are maintained by active transport mechanisms. GABA presides over one such mechanism by regulating the flow of chloride ions across the membrane. Barbiturates interact with the GABA receptor complex on the membrane to increase the flow of chloride through its channel, increasing the polarization and making the membrane less susceptible to depolarization. The chloride channel then seems to remain locked open for some time, causing prolonged depression of the cell.

A small dose of barbiturate produces a sense of calmness and relaxation that is comparable to a couple of pints of beer. It is an effective sleeping draught to start with, though likely to induce hangover effects and sluggishness the following day. Rapidly diminishing effectiveness following on regular use (tolerance) may lead to an escalation of intake to the point that depression of breathing becomes a risk. The ratio between therapeutic and toxic doses is small, and clumsiness, slurred speech, and unsteady gait may become apparent at quite modest doses. People taking barbiturates are notoriously accident-prone.

The barely detectable depression of breathing which occurs even at therapeutic doses makes the user vulnerable to bronchitis and even pneumonia. Otherwise, the effects of barbiturates are for all practical purposes confined to the nervous system, except in overdose.

Apart from the use of short-acting compounds in anaesthesia, the only current medical indications in Britain are for epilepsy, and 'severe intractable insomnia in patients already taking barbiturates'. Available preparations are amylobarbitone (Amytal), butobarbitone (Soneryl), quinalbarbitone (Seconal), and a mixture of amylobarbitone and quinalbarbitone (Tuinal).

There are virtually no statistics available on today's recreational use of barbiturates. Thirty-eight per cent of heroin addicts presenting recently to a drug dependency unit in Oxford said they had used barbiturates in the past two years, but none would admit to using them currently. A survey published in 1987 suggested that 15 per cent of addicts used barbiturates regularly, and more than half of these people injected them on some occasions.

They are not easily accessible to street drug users because only limited quantities are now manufactured, but are still prized as the drug of first choice by a small minority of oblivion seekers. Many of these people will be prepared to grind up tablets or dissolve the contents of capsules and inject the resulting sludge into a vein. Pill-injecting is a risky business at the best of times, but with barbiturates it is particularly dangerous (see p. 115). Others will swallow them if they happen to become available to mitigate the less welcome effects of stimulants, fend off opiate withdrawal symptoms, or get some hard-to-come-by sleep at the end of a day's hustling. Heroin addicts sometimes inject barbiturate unknowingly when dealers cut it into their 'gear' to give inferior material a bit more punch.

It is the medium-acting drugs that are most favoured on the black market (pento-, amylo-, quinal-, butobarbitone). Alcoholics have traditionally used these to fend off withdrawal symptoms. The combination of barbiturate and amphetamine has an appeal which transcends the current lack of a ready-formulated product. It seems that the stimulant's effect on energy, confidence, and strength, and the depressant's suppression of fear and anxiety combine to produce a particularly desirable high. The lack of availability of barbiturates makes it much more likely that the infinitely safer benzodiazepines will fulfil this role, although there is an unfortunate trend towards opiate use as an alternative.

Mild toxicity, often apparent at quite modest doses, is characterized by clumsiness, quivering pupils (nystagmus), slurred speech, and emotional instability. Alcohol and other brain depressants will potentiate these toxic effects. Minor side-effects include skin rashes, growth of hair in the wrong places, and swelling of the gums. As mentioned above, larger doses may result in depression of breathing, pneumonia, inefficiency of the heart and circulation, and kidney failure. Any of these can have a fatal outcome. Confusion may progress to delirium or coma, sometimes after a lag period of several days.

Long-term users often exist in a state of chronic impairment with symptoms of mild toxicity ebbing and flowing. Such people will exhibit unpredictable emotional states, switching between euphoria, depression, or irritability with little warning or provocation. Behaviour may be disinhibited and impulsive, and if influenced by a feeling of persecution dangerous aggression can result. Social deterioration is often apparent.

When tablets or the contents of capsules are ground up, mixed in water and injected into a vein, the resulting suspension is highly irritant. Local tissue damage may be intense, and injury to the vein, or even worse to the nearby artery or nerve, can threaten the viability of the limb. Because of the rapid destruction of superficial veins, intravenous users may be forced to go for the large veins in the groin or neck, with potentially disastrous consequences.

Sudden withdrawal of barbiturates will produce anxiety, agitation, trembling, stomach cramps, and sleeplessness. The sufferer may vomit, develop circulatory disturbances, or a fever. Without treatment, disorientation may be accompanied by false or illogical beliefs and hallucinations. There is a serious risk of fits occurring, and these can be very difficult to bring under control. Because of the potential seriousness of the barbiturate withdrawal syndrome, the procedure should be conducted in hospital or under close medical supervision.

There are a number of drugs which are not actually barbiturates but are very similar in their effects.

Ethchlorvynol (Placidyl) is a tranquillizer with a characteristic mint-like aftertaste, whose action comes on about 30 minutes after swallowing it, peaks at 90 minutes, and goes on working for between four and six hours. A hangover the next day is the usual sequel when it is used as a sleeping pill, and facial numbness is not uncommon. The profile of unwanted effects is very similar to that of the barbiturates.

Glutethimide (Doriden) enters the brain with particular rapidity, but is broken down in the body rather slowly. Apart from the usual barbiturate-like effects, it blocks the chemical messenger acetycholine quite powerfully. In the graphic phraseology of a 1940s article, this can make you '... hot as a hare, blind as a bat, dry as a bone and mad as a hen'. More prosaically, it can also cause loss of calcium from the bones or anaemia.

Meprobamate (Miltown) is an effective muscle relaxant and tranquillizer, and in 1957 became the most widely-used sleeping pill in the US because of effective marketing as a safer alternative to the barbiturates. It soon became a great favourite with opiate addicts for intravenous use. Its popularity waned as its true toxic potential emerged: fatalities were reported after as few as 15 tablets.

Although *methaqualone* (Qaalude) was withdrawn from legal supply many years ago, it is still highly prized on the black market. It is manufactured illegally in South America and elsewhere, and is usually available on the streets of London and New York if you have the right contacts. It is famed for its ability to produce a dreamy 'dissociative high' without sedation, an effect similar to that described following heroin. Unfortunately, this state is not compatible with taking good care of yourself and the drug has been associated with a number of deaths resulting from lack of attention, clumsiness, or disinhibition. There is also the possibility of rare but serious toxic effects including nerve damage, life-threatening anaemias, and fits.

Benzodiazepines

The benzodiazepines are phenomenally successful drugs which are prescribed in vast quantities throughout the world. In 1977, more than 8000 tons were consumed in North America alone. They form the major part of a worldwide anti-anxiety and anti-depression drug market worth more than $2 billion. Most people who take them do so perfectly legally, but there is a wide overlap between the licit and illicit arenas. There are no illicit manufacturers, the extensive black market being supplied entirely by diversion from legal supplies. It is legal to possess benzodiazepines without a prescription, but illegal to sell them or even give them away.

Since the middle 1970s, there has been growing concern at the prevalence of benzodiazepine dependence and misuse, and questions as to their continuing efficacy when used long-term. This has led to a gradual decline in prescriptions, a process which will no doubt gather momentum as more law-suits are brought against doctors and pharmaceutical companies by individuals who feel they have been turned into addicts. Although many doctors now limit their prescribing to carefully selected patients for no more than a few weeks, there are still those who argue that the reaction against the benzodiazepines has been excessive. They draw attention to the impressive safety profile on the one hand, and the high prevalence of anxiety disorders (around 10 per cent of the North American population) on the other. They point out that alternative medication for these people is less safe, and that non-drug treatments may be unacceptable, unavailable, or ineffective. There is some epidemiological evidence that when benzodiazepine availability is reduced, self-medication with more toxic depressants, such as alcohol, increases.

Extensive research in the fifties fuelled by growing awareness of the toxicity of barbiturates led to the synthesis of chlordiazepoxide in 1957. This was shown to have a 'taming' effect in animals and increased exploratory behaviour, eating, and drinking which had been suppressed by new or frightening experimental conditions. It was unveiled as Librium in 1960, and was soon followed by diazepam (Valium) in 1963. Both products were marketed aggressively by their manufacturer, Hoffman La Roche, and prescription sales accelerated rapidly as more variations on the theme appeared and the barbiturates were swept aside.

Repeat prescriptions became commonplace, a process aggravated in Britain by the fact that the vast majority of benzodiazepines were and are prescribed by hard-pressed general practitioners who are daily confronted by men and women attending their surgeries to demand treatment for sleeplessness and worry. The drugs reached their zenith in the mid-seventies, at which time well over 70 million prescriptions were issued yearly worldwide. Since then, a slow decline has been noted, more recently reinforced by limited prescribing lists established by District Health Authorities so as to reduce the local NHS drug bill, recommendations from such bodies as the Committee on Safety of Medicines, and publicity campaigns in the media. Even so, the prevalence of use remains impressive. Averaging across Western nations, around 15 per cent of adults use benzodiazepines each year, and 3 per cent use them daily for a year or more. The typical long-term user is a woman aged over 50 with multiple, chronic health problems.

All benzodiazepines are well absorbed by mouth, and most are broken down in the liver to active and inactive metabolites, some of which may be extremely long-acting. There is a wide range of compounds on the market, and the use to which they are put depends to a large extent upon how rapidly they disappear from the blood-stream. They are categorized on this basis as short-acting (temazepam, oxazepam, loprazolam), medium-acting (lorazepam, alprazolam), and long-acting (nitrazepam, flunitrazepam, chlordiazepoxide, diazepam, clobazam, flurazepam, ketazolam). Unfortunately, the blood level does not always reflect brain levels and so may be misleading. There is a complicated chemical interrelation between the drugs. For example, one of the metabolites of diazepam is temazepam, which in turn has oxazepam as one of its breakdown products.

In 1975, it was realized that benzodiazepines work by enhancing the effects of the brain's most important inhibitory chemical messenger (neurotransmitter), gamma amino butyric acid (GABA). In 1977, a specific receptor site for benzodiazepines was discovered to be part of the GABA complex on cell membranes of nerve cells. This suggests that there must be some naturally occurring substance in the body which has a similar effect to these drugs. Barbiturates, and possibly alcohol, have an effect at the same site.

The way that GABA inhibits nervous activity by controlling the flow of chloride through a specific channel in the cell membrane has been described above in connection with the barbiturates. Since GABA is very widely distributed throughout the brain and spinal cord, the effects of the benzodiazepines are diffuse throughout the nervous system. Inhibitory transmission in the spinal cord results in a muscle relaxant effect peripherally.

Benzodiazepines are not particularly powerful in animal dependency models, but greatly reduce the suppressive effects of punishment on animal behaviour. There is no evidence of nerve damage in long-term dosing in animals or humans. Like most sleeping pills, they alter the pattern of rapid eye movement (REM) and 'slow wave' phases of sleep. The pattern bounces back in the opposite direction when they are discontinued, giving rise to 'rebound insomnia'.

Benzodiazepines are still widely prescribed for sleeplessness and anxiety. Other indications include supplementation of anaesthesia, muscle spasm or cramps, and epilepsy. From the pharmacological point of view, they are remarkably benign drugs with virtually no adverse effects outside the nervous system. Breathing can be depressed by enormous doses, but a fatal outcome from benzodiazepine alone is extraordinarily rare. In combination with other brain depressants such as alcohol, the danger is much greater. They cross the placenta and do seem to be associated with a small increase in the incidence of birth defects. When taken late in pregnancy, they may produce the 'floppy infant syndrome'. As with all drugs, they should be avoided in pregnancy if at all possible, especially in the first three months. They enter breast milk and can make the baby sleepy.

Non-medical use of benzodiazepines has been documented since the sixties, and in 1985 they were used in this way by four per cent of the US population. In other countries, surveys have indicated that in some areas as many as 10–15 per cent of adolescents have used them recreationally (see footnote page 3). Most of this use is by mouth, but tablets and the contents of capsules are occasionally ground up and snorted into the nose. Intravenous use is rare except among those already accustomed to using other drugs by this route.

Amongst attenders at drug dependency units (DDUs), benzodiazepines are rarely drugs of first choice, but a large majority use them from time to time, often in huge doses. A survey in

Oxford indicated that 81 per cent of heroin addicts had used them in the previous year, and 27 per cent were using them daily at presentation. Other surveys have suggested an even higher prevalence than this. The price on the black market is variable, but always cheap: even on a bad day, £1 will buy three or four diazepam or temazepam tablets.

Awareness that the intravenous use of benzodiazepines, especially temazepam, was becoming a significant problem grew from the mid 1980s. It tended to be focused in particular areas of the country, and did not necessarily reflect a shortage of more conventional merchandise such as heroin. By the end of the eighties, the practice had become more widespread, with up to a third of injecting addicts admitting to it. One survey suggested that as many as 70 per cent of injecting drug users had injected temazepam at some time. The manufacturers of temazepam did not welcome the publicity they were getting, and in 1989 reformulated their liquid capsules as a hard gel in an attempt to make them uninjectable. Determined addicts found they could overcome this hurdle by warming the gel and using larger needles. Not surprisingly, this practice carries with it terrible health risks, since the gel tends to harden up again in the bloodstream and can easily block vessels.

A recent survey suggested that about half the benzodiazepines obtained on the black market were taken with a view to obtaining a 'buzz' or high. Other important reasons included the need for sleep or rest, to limit the unwanted effects of stimulants or hallucinogens, to gain confidence to carry out criminal activities or other stressful tasks, and to boost the effects of opiates, especially heroin or buprenorphine (Temgesic). Almost half of those questioned thought tranquillizers made them more violent and disinhibited, and many thought their memory was impaired by them.

A lot of people seem to find that a handful of valium or temazepam takes away all the worries of street life for a while. Adolescents sometimes take them as an alternative to 'Special Brew' or solvents. At modest doses, the effects are very similar to those of alcohol, but increasing resistance to the euphoriant and sedative effects mean that some people get up to massive amounts with no visible signs of intoxication: the equivalent of twenty or thirty times the manufacturer's recommended maximum dose of diazepam taken daily is by no means unusual. Such an intake is quite simple to maintain because the drugs are so cheap and easy to come by. Tablets can be obtained from enterprising pensioners whose repeat prescriptions for insomnia generate a useful supplement to their income, from alcoholics who prefer alcohol to the librium or chloral they are prescribed for withdrawal symptoms, or from the medicine cupboard of a hard-pressed parent. As prescribing has slowly declined, so the seizure rate by police of illegal stockpiles has increased in recent years. Benzodiazepines form part of the regular currency and barter of street life.

Physical risks associated with the benzodiazepines are mainly by-products of depressed mental functioning, for example clumsiness, forgetfulness, or disinhibition leading to self-neglect or accidents. As mentioned above, depression of breathing can occur with massive doses, or more easily if other depressant drugs are taken at the same time. Very rarely, abnormalities in liver function crop up which are reversible if the drugs are withdrawn.

Injecting benzodiazepines is associated with all the usual risks of this practice: septicaemia, hepatitis, HIV, and other infections; blockage of veins, damage to nearby arteries or nerves which can lead to loss of the limb; abscesses or ulcers at the injection site. The newly formulated hard gel temazepam 'eggs' are particularly likely to damage veins or lead to blockage of vessels. Injecting tranquillizers seems to have a particular association with a self-destructive, chaotic, deviant life-style. People whose lives have taken this course are more likely to share injecting equipment, have unprotected sex, and take more accidental and deliberate overdoses.

Many people take prescribed benzodiazepines for months or even years and find it hard to contemplate doing without them. Is it worth the bother of battling to give up what is after all a cheap and relatively non-toxic drug? The answer is that for some people, chronic use does carry definite disadvantages. There can be problems with concentration and memory; depression, tiredness, and apathy; clumsiness and accident-proneness which has particular significance for drivers; disinhibition and instability of mood. The ability to cope with everyday problems may be impaired. From time to time, withdrawal symptoms may emerge due to altered sensitivity to the regular dose. The elderly are particularly prone to all these effects. With toxic doses, confusion and memory deficits worsen and control over bodily activity deteriorates.

Physical dependence on benzodiazepines is now clearly recognized and may become apparent after as little as six weeks or so of regular ingestion. Symptoms overlap with those related to psychological dependence (see Chapter 10), and I will consider them together. The withdrawal syndrome has to be distinguished from re-emergence of the pre-existing anxiety state. The proportion of people reported to experience a withdrawal syndrome after coming off long-term benzodiazepines ranges from 20–90 per cent, but most studies suggest that it will occur in more than half. It is difficult to predict with any confidence who is likely to suffer, but those people who might be described as neurotic are very vulnerable, as are those with a history of dependency on other substances. Unsurprisingly, it is more likely when larger doses are taken over longer periods. Abrupt termination results in more symptoms than tapering the dose off gradually. Shorter-acting drugs cause more intense withdrawal symptoms, and are associated with greater rates of relapse. Relapse may occur in up to half of those trying to give up, but people who successfully remain abstinent for five weeks or more report lower levels of anxiety than when they were still taking the drugs.

Onset of withdrawal symptoms occurs within a day or two of cessation in the case of short-acting compounds such as temazepam or lorazepam, but can be delayed up to a week or so following the more slowly-metabolized drugs such as diazepam. Patients will complain of anxiety, restlessness and irritability, heightened sensitivity to noise and other environmental stimuli, twitches and shakiness, weakness and lack of motivation or energy, stomach cramps, headaches, poor concentration or confusion, sleeplessness, and a strong desire for tranquillizers.

These symptoms vary greatly in severity, and usually last for about two weeks after the last tablet has been taken. Fits will occur in up to 20 per cent of patients withdrawn abruptly from substantial doses.

Anyone taking benzodiazepines regularly for longer than three weeks or so risks developing both physical and psychological dependence. This is particularly likely to occur in people suffering chronic stress, or those with rather a passive nature, a past history of dependence on other drugs or alcohol, poor coping skills, impulsivity, or subject to rapid and unpredictable mood swings.

There are a number of drugs which are not actually benzodiazepines but have a very similar effect.

Chloral hydrate (or chloral betane, 'Welldorm') is marketed as a sleeping tablet. It is rapidly absorbed by mouth and the blood level halves in around six hours. It has a characteristic bitter taste and is irritating to the stomach, and so should be avoided by anyone with gastritis or a history of peptic ulcer. It is also dangerous for those with severe heart, liver, or kidney disease. Breast-feeding mothers who take it will make their babies sleepy. Chloral is widely available on the black market.

Chlormethiazole (Heminevrin) is another sleeping pill, and is structurally related to vitamin B1 (thiamine). It is broken down rapidly in the body, with blood levels halving every four hours. It has similar contraindications to chloral, and should be avoided during pregnancy and by breast-feeding mothers. It seems to cause sneezing and itchy, red eyes. It is often prescribed to alcoholics to relieve their withdrawal symptoms, and potentiates the toxic effects of alcohol in those who continue to drink. Chlormethiazole spills liberally on to the black market from this and other sources.

9
Heroin and the opiates

The opium poppy has always provided a source of comfort to mankind. For thousands of years, ordinary people have eased their lives through the power opium has to soothe and sedate, and this folk-knowledge found its expression in the great pharmacopoeias of the Egyptians, Sumerians, Greeks, Persians, and Romans.

Each age, including our own, has spun its myths around the drug. Theophrastus was defining its medical applications in 320 BC but the ancient Greeks also knew that when the goddess Demeter discovered the poppy's secret she forgot all her sorrows. What else could Helen of Troy's *nepenthe* have been but opium?

Today, the need and desire for opium and its derivatives remains as great as ever. Strict international legal control of supply has opened the door to a worldwide multi-billion dollar crime network whose malign influence has seeped into every part of society, and whose windfall profits are built upon the exploitation and misery of millions.

Although opiates remain central to pain control in Western medicine, thousands of hospital patients and others suffer needlessly because of the unspoken fear of inducing dependency. Heroin addiction has become a terrible scourge of our times, but has this resulted primarily from the properties of the drug itself or the way that society has reacted to it?

From its roots in the ancient world, knowledge of opium was borne along the arteries of commerce. Arab traders introduced it into India in the seventh century AD, and to the shores of China a couple of hundred years later. Another messenger was the common soldier. Opium found its way to Europe in the pouches of returning Crusaders at the start of the new millenium.

Paracelsus (1490–1541), the first European to recommend the targeting of specific symptoms with specific remedies (for which he has been called the 'father of modern pharmacology') invented a mixture of opium, alcohol, and spices which he called laudanum ('worthy of praise'). With minor modifications to the formula, laudanum was to soothe and torment millions of Europeans over the next 400 years.

Although the possibility of addiction was recognized from the earliest times, nobody seems to have worried much about it until relatively recently. Perhaps it was in eighteenth-century China that the first high-profile problems can be detected. At this time, the British thirst for quality tea was already highly developed and the Chinese produced the best product available. Unfortunately, Britain was rather short of anything that these sophisticated people might wish to trade for it. The solution to this dilemma came in 1773 with the conquest of Bengal, which gave the British a monopoly in Indian opium. The East India Company set about exporting this in large amounts to China, getting round quota problems with piratical zeal.

Up to this time, non-medicinal use of opium by the Chinese was not prominent, but opium smoking seems to have caught on in a big way and paying for these huge imports soon became a serious embarrassment to the Chinese exchequer. The Emperor made the first of many attempts to ban the habit and limit imports in 1799 but in 1839, after decades of helpless frustration, he was driven to take more active steps with the seizure and destruction of large amounts of the drug in Southern China. Lord Palmerston promptly despatched sixteen warships to besiege Nanjing, and what later became known as the First Opium War was underway. It ended in 1842 with massive concessions to the British, including the ceding of Hong Kong island as a colony 'in perpetuity'. Further fighting took place between 1856 and 1860, but this simply led to more humiliation for the Chinese. By this time, it is estimated that there were between 15 and 20 million opium addicts in the country.

Opium was eulogized in the European medical textbooks of the eighteenth and early nineteenth centuries. Not only was it manifestly effective in relieving a tremendous range of symptoms, it was even recommended for use by those in the pink of health because of its ability to 'optimise the internal equilibrium of the human body'. In England, it occupied a position as a popular home remedy rather similar to that of aspirin today, with consumption peaking at eleven pounds per thousand population in the middle of the nineteenth century.

Use of opium for pleasure or the relief of fears and anxiety was widespread throughout society. Every one knows of the famous writers who were enslaved or bewitched by it: Thomas de Quincy, S. T. Coleridge, Wilkie Collins, Charles Dickens, Elizabeth Barret Browning, and many others. The ability of this and other drugs to enhance creativity remains a matter of vigorous debate to this day. Less well publicized were those politicians and members of the professions who were also awash with laudanum. But it is important to realize that opium was well within the financial reach of the poorest in the land. There were agricultural workers in the Norfolk fens who would have regarded a mug of ale without a knob of opium in it with as much enthusiasm as a modern business executive might view a gin and tonic without the ice. Of course, one could always resort to a bowl of poppy-head tea; a selection of poppy heads was available in some British pharmacies right through to the 1950s.

What were the problems associated with this completely unregulated free availability of opium? The risk of dependency was certainly recognized, but those who succumbed were regarded with rather amused tolerance for the most part. It is interesting that this tolerance was generally restricted to opium 'eaters' (drinkers would be a more appropriate term, since laudanum or patent medicines were by far the commonest form in which the drug was available). Opium smoking seems always to have been regarded as a rather vile alien indulgence.

Opium-containing medicines were widely used as child-calmers, and the mortality associated with this became a major concern in the middle of the nineteenth century. Most at risk were the small children of working women from the new industrial classes. Inflexible working hours meant leaving children for long periods unattended. Over-sedation in such circumstances could easily have fatal consequences, and recorded cases of this sort (likely to be a considerable underestimate) were running at around 80 per year. In the general population, fatal poisoning with narcotics made up about a third of all poisoning deaths.

This growing concern, along with a certain measure of professional self-interest, led to the introduction of the Pharmacy Act in 1868, which restricted sale of opium and its derivatives to registered chemists but had little or no impact on the volume of usage. This only began to decline significantly towards the end of the century as the public perception of the drug became less favourable. This period saw a number of law-suits against patent medicine manufacturers, and a rapid decline in the popularity of these products. The Act was made rather more restrictive in 1908.

Opium was brought to North America by the early European settlers, and quickly became familiar to most households as in Britain. With characteristic energy and enthusiasm, entrepreneurs were quick to see the commercial possibilities; America's first opium-containing patent medicine appeared as early as 1796. The dangers of excessive use were recognized just as quickly and were laid out in some detail in the Dispensatory of 1818 but, as in Europe, the drug soon became absolutely central to mainstream medicine. A prominent American physician (J. Periera) wrote in 1846 'Opium is undoubtedly the most important and valuable remedy of the whole *Materia Medica*'.

Laudanum could be bought for a few cents an ounce, and Americans took to patent medicines even more enthusiastically than the Europeans had. The foundations of some colossal modern fortunes were laid in those unrestricted times, and by the end of the century more than 50 000 patent medicines were on the market. Poppies were cultivated throughout the land, and business was booming. Opiate addiction at that time was said to be much more common in women, possibly because social restrictions limited their access to alcohol.

Two nineteenth century advances laid the foundation for the scale of problems we face today. First, the active ingredients of opium were identified, paving the way to the creation of immensely potent synthetic opioids. Second, the technique of intravenous injection was perfected.

In 1805, a German pharmacist's apprentice called Sertuerner isolated a chemical from raw opium which he initially called *principum somniferum*. Later, he hit upon the snappier name morphine, after Morpheus, Greek god of dreams. This chemical was found to make up 10 per cent of the weight of opium, and to be ten times more potent. Isolation of codeine followed in 1832, thebaine in 1833, and papaverine in 1868. In 1874, diacetyl morphine was synthesized but was not marketed by Bayer until 1898. For many years after this it was considered a highly effective remedy for morphine addiction, and its tradename was derived from the German word *heroisch*, meaning 'powerful'.

The architect Christopher Wren was one of the first to experiment with injecting substances into the bodies of both animals and humans. In the early days, a blob of material would be placed on the skin, and repeatedly jabbed with a quill or something similar. Then someone hit upon the idea of inserting a hollow tube into a vein and later in the seventeenth century there were even attempts at blood transfusions. The outcome of these early experiments was extremely disappointing, particularly for the recipients. Interest was revived in the nineteenth century, and in 1858 a Scottish surgeon adapted an instrument which had been designed to drain blood from birth marks to deliver a dose of morphine as closely as possible to the point of pain. This device became known as the hypodermic syringe. Administering drugs by injection very rapidly transformed medical practice, and the face of addiction too.

Despite the widespread use of opium and its derivatives in nineteenth-century America, the practice of opium smoking remained socially unacceptable as in Britain. This was largely due to the negative attitude of the native population towards the Chinese immigrants who had been brought over in large numbers to work on the railroads in the 1850s. When a deep recession arrived soon after this with large-scale unemployment in its wake, these hard-working foreigners were deeply resented. Their habit of opium smoking became symbolic of a perceived degeneracy and debauchery. In an age of almost totally unrestrained drug use, opium smoking was made illegal in San Francisco in 1875.

It is often said that the American Civil War with its widespread misery and suffering gave a huge boost to opiate usage, and it is true that morphine addiction became known at the time as 'The soldier's disease'. Certainly, vast amounts of opium and morphine were issued to soldiers on both sides in their field packs, but the effect on addiction prevalence remains controversial.

Attempts to regulate the booming patent medicine business and its outrageous marketing techniques were very slow in coming to the US. Not until 1906 did the Federal Pure Food and Drug Act require the ingredients to be specified on the label. Public concern mounted rapidly in the early years of the twentieth century, and action to stem misuse of opiates suddenly became a vote-winner for politicians. In 1909, importation of smoking opium became illegal, and America was instrumental in bringing about a series of international conferences on narcotic abuse, first in Shanghai then at The Hague. In 1912, it was agreed that the participating countries (by then numbering more than 30) should go away and enact domestic legislation to restrict the use of opiates to medical indications on prescription from a doctor.

Accordingly, the Harrison Act entered the statute books in 1914. There then followed a lengthy debate as to whether the maintenance prescription of opiates to confirmed addicts by a physician constituted acceptable medical practice. In 1919, the Supreme Court ruled that it did not, thereby setting in motion a penal response to drug addiction which remained firmly in place until the AIDS epidemic forced a change of policy in the 1980s.

The public alarm which made possible these legal and political moves is understandable given the size of the problem at the time; it was conservatively estimated that there were at least 275 000 opiate addicts in the US in 1920. In Britain, the situation seemed to be very different. Despite the massive use of opium and morphine throughout the previous century, opiate addiction was not a particularly visible problem in the early years of the twentieth century. The British were in no great hurry to fulfil the agreement to legislate that they had made at The Hague. This may not have been totally unconnected with the fact that Britain produced most of the world's supply of morphine at that time.

This laissez-faire approach was quickly abandoned during the First World War when it became clear that British soldiers on leave in London were understandably keen to get hold of drugs which might make life in the trenches a bit more acceptable. There was a roaring London black market in cocaine, but the fashionable shops were getting in on the act too. Harrods, for example, sold morphine and cocaine kits complete with syringe and spare needles labelled, 'A Useful Present for Friends at the Front'. An order came from army big-wigs forbidding the supply of drugs to soldiers on active service lest their efficiency or will to fight be impaired, but this merely fuelled the black market. The order was then given teeth by the introduction of Defence of the Realm Act Regulation 40 B in June 1916. This primarily targeted cocaine, but restricted opium as well for good measure.

After the war, there were some lurid stories in the newspapers of drug-soaked debaucheries in the night clubs of London's West End. The sordid death of a well-known actress and the subsequent high-profile prosecution of her dealer gave the Home Office the opportunity to press for immediate civil legislation. This was unsuccessfully resisted by the fledgling Ministry of Health, and in 1920 the DORA 40 B was enacted with many extra restrictions as the first Dangerous Drugs Act.

For a few years, it looked as though the British response to drug problems would follow the penal route chosen by the Americans. But in 1926 a Government-sponsored committee chaired by the President of the Royal College of Physicians was asked to examine the logic and practicalities of this course. The Rolleston Committee recommended that addiction should be regarded as an illness requiring treatment, rather than a crime demanding punishment. Long-term maintenance on a prescription of opiates was recommended for those addicts who were unable to give up but who could lead a 'useful life' when so maintained. The 'British system' for the treatment of opiate addiction thus came into being.

Whether this admirably humane approach would have found support if addiction had been other than a thoroughly middle-class affair has often been questioned. It is certainly true that there were very few working-class opiate addicts known to doctors in the Britain of the twenties and thirties. Most came from the professional classes, and for some reason injectable morphine use was particularly prevalent among respectable housewives. This very stable, almost placid picture remained in place right through to the 1960s when it was suddenly and rudely disturbed.

In the United States after the First World War, the situation was very different. The visible plight of the large numbers of opiate addicts led many doctors to hope that the Harrison Act would not be vigorously enforced. Addiction clinics were set up, and large numbers of patients maintained. The Narcotics Division within the Bureau of Prohibition had other plans. In 1922, agents began a clamp-down on the clinics, and initiated prosecutions against addicts, doctors, and pharmacists. At first, many physicians were prepared to fight these prosecutions in the courts, but after a number of careers had been ruined the heart went out of the fight and by 1925 the medical profession had withdrawn from the addiction field. By this time heroin had also been banned from use in medicine, and its legal manufacture in the US ceased completely.

Half a million opiate addicts were thus abruptly cut off from their drug supply. The racketeers who had already grown fat on the proceeds of bootleg whisky couldn't believe their luck. Supplies of opium were cheap and easy to come by, morphine and heroin were easy to manufacture in backstreet laboratories. The bad guys were not slow to respond to the demands of their desperate customers, and the black market in drugs was soon consolidated nationwide. When prohibition of alcohol was repealed in 1933, the mobsters were unconcerned. By this time, undreamed-of profits were being achieved in the most lucrative criminal enterprise of all time. Since the twenties, no world event has ever seriously interfered with the flow of narcotics into North America.

American addiction rates probably fell during the Second World War and for a time thereafter, but rose steadily in the fifties as the Mafia reopened trade routes from Turkey and South-east Asia. In particular, there was tremendous demand from urban Black and Puerto Rican communities which the mob was only too pleased to meet. The penal response became more savage in an attempt to reverse the trend. The 1951 Bogges Act had already defined minimum mandatory jail terms for possession with no possibility of parole in some cases. In 1956, the Narcotic Control Act doubled maximum sentences, and made it possible to award dealers the death penalty.

By this time, it was becoming noticeable that the average age of American addicts was falling steadily: in the 1930s, less than 20 per cent of male addicts were under 30 years of age, but by 1962 more than half were in their twenties.

Observing this trend with concern, the British government decided to review the state of affairs in the UK. The Brain Committee's first report in 1961 was immensely reassuring. Nothing had changed since the twenties, just a small number of well-behaved professional types plodding along quietly on their scripts, and a handful of bohemian addicts in Soho and West Kensington.

Barely was the ink dry on the parchment when it became obvious to anyone visiting a jazz club or a coffee bar in West London that things were changing fast. The Brain Committee was hurriedly reconvened, and its second report had to concede that there was a growing problem requiring drastic practical measures. The recommendations included restrictions on who could prescribe heroin and cocaine, compulsory notification of opiate and cocaine addicts by doctors to the Home Office, and the establishment of special Drug Dependency Units (DDUs). The Misuse of Drugs Act was passed through Parliament in 1970.

Why had the drug culture suddenly exploded? Economic stability meant money in people's pockets, better international communications were leading to the emergence of a powerful and independent youth culture, and a number of established addicts escaping the repressive policies of their own countries arrived in England to take advantage of the British System. This hitherto relaxed and restricted philosophy was being reinterpreted by a small number of London doctors in a way which was to have a profound effect on the local drug scene.

In an attempt to keep their demanding patients out of the clutches of criminal suppliers who were imagined to be lurking on every street corner, or more occasionally for simple monetary reward, these doctors were prepared to prescribe truly awesome quantities of heroin, cocaine, or injectable amphetamine to their young clients. One infamous doctor prescribed at least 600 000 tablets of heroin (for injection) to addicts in 1962. It was apparently not unusual for her to give an addict as many as 900 tablets on a single occasion, and be prepared to replace them a few days later if they were reported 'lost'. She went on a lecture tour of Canada to describe her philosophy; not surprisingly, a number of Canadians decided to return to the Old Country. Some 'drug-doctors' held their surgeries in pubs or station buffets. When the restrictions on heroin prescribing were introduced, some were undeterred or simply switched their addicts to other drugs. One who was later struck off the medical register is said to have prescribed more than 25 000 ampoules of methamphetamine in one month alone. By 1967, there were 2000 known heroin addicts in the UK.

The irony is that there was virtually no criminally organized black-market in Britain at that time. Almost all drug seizures up to the end of the sixties were of pharmaceutical heroin, cocaine, or amphetamine—overspill from this orgy of prescribing.

In both the UK and America there seems to have been a further upsurge of heroin use as the hippy movement degenerated, in the wake of an outbreak of methamphetamine injecting which swept the big cities. Heroin was one way of coming down from this switchback ride of a drug. A sordid scene centred on Piccadilly Circus in London, and other places conveniently close to all-night chemists. Two peaks of prevalence seem to have occurred from 1968–72 and 1974–76: at these times, the drug was unusually plentiful, of relatively high quality, and cheap. Heroin use became endemic in provincial cities during the seventies and eighties. Street dealing seems to have become associated with more and more violence. It is an unwelcome and recent change in Britain that some inner-city areas have become virtual no-go areas to law-abiding citizens at certain times.

In periods of relatively short supply, 'designer drugs' tend to become more prominent on the streets, especially in America. These are chemical variations on a theme, and the opioid designer drugs have frequently been based upon fentanyl, a synthetic compound 200 times more potent than morphine. Alpha-methyl fentanyl appeared on the streets in 1979 as 'China White', so-called because of its similarity in appearance to Chinese heroin of a particular type. Within ten years, back-street chemists had come up with at least ten variations of this basic drug, and it is estimated that a quarter of California's heroin addicts used a fentanyl variant at some time. One potential advantage for the user of these drugs is that they do not show up in standard laboratory tests for heroin. The high potency of China White caught some people on the hop: at least a hundred overdose deaths were associated with it.

Analogues of pethidine (meperidine) also crop up from time to time. Between 1982 and 1985, methyl-phenyl-propionoxy-piperidine (MPPP) and another drug with a long chemical name (MPTP) were marketed as 'synthetic heroin'. Unfortunately, there was a toxic by-product of manufacture which resulted in more than 500 people being permanently brain-damaged, yet another example of the lack of quality control in the world of criminal chemistry.

International gangsters continue to enjoy the billion-dollar bonanza that illicit heroin provides. Mafia, Camorra, Triads, Yardies, even the Hell's Angels, there's more than enough profit for all of them. The source material comes from the countries which form the opium crescent—Turkey, Iran, Afghanistan, Pakistan, India, Nepal, Bangladesh, Burma, Thailand, Laos, and other countries of South-east Asia. Mexico also contributes, especially to the US scene. Sources, routes of supply, and location of refinement plants change according to the exigencies of law enforcement successes, weather, demand, and the activities of rival gangsters.

Raw opium is obtained from the seed capsule of the poppy, *Papaver somniferum*. When the capsule is slit after the petals have fallen, a white juice oozes out which soon thickens and darkens to a tar-like substance with a bitter taste and pungent smell. This is scraped off, then rolled into balls or squashed into bricks. Active ingredients make up about a quarter of this raw material by weight. Opium for smoking is easily prepared by repeatedly boiling and filtering this stuff until it takes on the consistency of treacle.

Terminology for the derivatives of opium can be a bit confusing. Narcotic (derived from the Greek word meaning stupor) is a misleading term because it does not reflect the principal activity of the drug. Some people use the term opiate for those drugs directly obtained from opium, and opioid for all other morphine-like compounds. For simplicity, I will use the latter term to represent all drugs, natural or synthetic, with a morphine-like action.

The pharmacology of the opioids is now quite well understood. In the 1960s it had been suggested that they might bind to specific receptors in the nervous system in order to exert their effect, and several were soon identified. What was the function of these receptors in nature? The answer came with the discovery of the 'brain's own opiates' in 1975. It became clear that certain peptides (basic building blocks of protein made up of strings of amino acids) are produced by the body in response to pain and modulate its perception. They achieve this by acting as short-lived chemical messengers in the brain and elsewhere, modifying and shaping the effects of other important chemicals such as serotonin (5HT), dopamine, and noradrenaline. These peptides were given the collective title endorphins and several have now been isolated, including met- and leu-enkephalin, dynorphin, and the rather unimaginatively titled peptide-E.

The opioids, then, latch on to the places (receptors) on nerve cells where the endorphins would normally roost and influence the function of the nerve, either by altering the flow of chemicals across the cell membrane and hence its electrical polarity, or by modifying the release of other chemical messengers. The picture is further confused by the fact that different opioids vary in the way they affect the receptor. Some stimulate it and produce a positive effect (agonists); others sit on the receptor and block it without turning the system on (antagonists); a third group are capable of bringing about either of these outcomes (partial agonists). Sometimes, it is difficult to predict which effect will predominate in this third group. For example, if buprenorphine (third group) is given to a heroin (first group) addict, it may either intoxicate the person or induce a withdrawal reaction.

Opioids exert a mainly inhibitory effect upon the brain and nervous system, with a wide range of results. Perception of, and concern about, pain is reduced; the pupils of the eye are constricted; drowsiness or sleep is induced with larger doses; anxiety, panic, and fear are inhibited, as is the 'fight and flight' mechanism; the cough reflex is suppressed; breathing is depressed; and body temperature is reduced.

Sex hormones are reined in, whilst prolactin and growth hormone are increased. Peripherally, the blood vessels relax and blood pressure falls, the guts slow down, the muscles loosen and the sphincters tighten. In short, the organism slows up and a state akin to brief hibernation ensues.

Most opioids are not very effective when taken by mouth because the liver neutralizes most of the dose before it can reach the brain. An exception to this is methadone (synthesized in 1941 by German chemists) which also lasts much longer in the body. It takes 24 hours or more for the level in the blood to fall by half, compared with four hours for most other opioids. These properties have led to methadone being the most widely prescribed substitute opioid for addicts in treatment.

Some of the more commonly used opioids are listed below. The roughly equivalent dose to 10 mg morphine (a fairly modest pain-killing dose) is given in brackets: diamorphine (heroin, 4 mg); codeine (120 mg); hydromorphone (Dilaudid,

1.5 mg); hydrocodone (10 mg); pethidine/meperidine (Demerol, 100 mg); dextromoramide (Palfium, 7.5 mg); dipipanone (Diconal, 10 mg); methadone (10 mg); buprenorphine (Temgesic, 0.4 mg). The main uses of opioids in medicine today are in the treatment of severe pain, pre-medication for surgery, suppression of cough and diarrhoea, and for the alleviation of a particular form of heart failure in a hospital setting.

After an adequate dose by intravenous or intramuscular injection (at least 10 mg for an average-size adult), morphine is effective against pain within minutes and lasts for three or four hours. It is well known that many patients suffer needlessly in hospital because of a fear amongst the clinical staff of inducing addiction. In fact, this is vanishingly rare in patients treated for pain; approximately one in a hundred thousand treatments in one series.

Opioids are extraordinarily safe drugs in ordinary medical practice, except in overdose when depression of breathing may be life-threatening. If recognized, this can easily be reversed by injections of the opioid antagonist naloxone. Since this has a very short duration of action, it may have to be given repeatedly until the opioid has worn off. The danger of fatal respiratory depression is greatly increased if other brain depressants such as alcohol, tranquillizers, or antidepressants have been taken as well. People with lung complaints or liver disease are particularly vulnerable to this dangerous effect.

Repeated doses may be associated with increasing resistance to many effects of the drug (tolerance), necessitating larger doses to achieve the same result. Some effects, such as constipation and pupillary constriction, do not exhibit tolerance.

There is no evidence that opioids directly damage the unborn baby but they do cross the placenta, so neonates may be slow to breathe, or show withdrawal symptoms if the mother was taking them regularly. Breast-feeding mothers should bear in mind that these drugs enter breast milk and may sedate the infant.

Current estimates, and they are no more than estimates, are that around 500 000 Americans and between 100 000 and 150 000 Britons are dependent upon injected opioids (see footnote page 3). It is important to note that this is the number of *visible* opioid users. They are visible either because they have presented to doctors or others requesting help, or have come to the attention of the police; by definition, they are 'problem drug users'. Some research indicates that a larger number of opioid users remain invisible. As many as one per cent of the adult American population will admit to having used opiates recreationally at least once. One study suggests that as many as three million Americans may use opioids from time to time, only about 10 per cent of whom could be described as being addicted. These intermittent, non-dependent individuals have been called 'chippers'. It has been estimated that only 10–15 per cent of opioid addicts are in contact with the health services at any one time.

Opioid use is common among convicted criminals, most of whom have a record which antedates their drug use. One study revealed that 28 per cent of the inmates of a Scottish prison were intravenous drug users prior to their incarceration. Eight per cent had injected drugs at least once in prison and of these three-quarters had shared injecting equipment. Heroin does not seem difficult to come by in prison, but injecting equipment can be in short supply. Ex-prisoners will tell you that dozens of inmates will sometimes be sharing a single needle and syringe; a truly terrifying prospect in the HIV age.

The number of people asking for help with opioid addiction has increased steadily through the eighties: 61 689 presented to British drug dependency units in 1989 compared to 13 905 in 1985. It is a condition with a considerable mortality, with 2743 deaths in the US directly attributable to opioid abuse in 1989, which accounts for 38 per cent of all drug-related deaths (cocaine accounted for 50 per cent). The American Poisons Center has estimated a death-rate of 17 per thousand exposures, which seems rather a high figure.

Four times as many addicts were notified to the British Home Office in 1990 than was the case in 1980, though this may just indicate greater efficiency by doctors. The amount of heroin seized peaked in 1990 at over 600 kg, fell back to between 400 and 500 kg in 1991 and 1992, but was almost back to 600 in 1993. Average purity is around 40 per cent, and prices are quite static at about £80 per gram. American street heroin only costs about $80–100 a gram but it is very impure, containing as little as 3–5 per cent diamorphine.

Street heroin reaches the user by what amounts to a massive pyramid-selling operation. The importer has contact with a number of distributors who connect with people who will buy a few pounds in weight, who in turn sell to others in ounces. These ounce dealers are often users themselves. The next level consists of street users who can afford to buy several grams at a time and then parcel them out in smaller and smaller quantities right down to the lowly £10 bag or wrap. The drug becomes more adulterated (cut, 'stepped on') at almost every point of this journey. So most heroin users are also dealers at some level, even if the profit simply goes towards funding their own habit.

Substances which have been identified in street heroin ('gear') range from the relatively harmless—mannitol, glucose, lactose, and sucrose; the not-so-harmless (especially to the intravenous user)—caffeine, ephedrine; and the potentially lethal—talcum powder, flour, brick dust, Ajax, barbiturates, and strychnine.

Prepared opium for smoking is hard to come by in the West, and is usually presented as a material resembling treacle, or compressed into pills or sticks. A small number of patients receive pharmaceutical diamorphine on prescription as a white powder in 'dry amps', since diamorphine decays quite quickly in solution. These appear on the black market from time to time. Most addicts on prescription receive methadone as a green, rather sticky mixture, or less commonly in ampoules for injection. This will cost round £10 for 100 millilitres of mixture or £5 an amp on the street corner. Pharmaceutical morphine does not seem very popular with addicts unless formulated with the anti-vomiting drug cyclizine, but Palfium and Diconal tablets are highly sought-after. Various milder opioids all have some black-market value, and proprietary cough medicines containing small amounts of morphine or codeine come in handy to fend off withdrawal if street supplies run short. Buprenorphine is particularly fashionable in some parts of the country.

Street heroin usually consists of an off-white or brown powder, although it may occasionally be granular in constituency. It has a bitter taste, though this may be artificially heightened by the addition of quinine. The powder can be snorted up the nose, smoked in a cigarette or over heated foil ('chasing the dragon') or dissolved and injected ('fixed'). Most of the people who present to doctors or others with drug problems either inject or chase.

Chasing involves placing some powder on a piece of silver paper, applying a flame beneath, and inhaling the fumes through another piece of silver foil rolled into a tube. The rush (sudden wave of euphoria) achieved by this method is almost as good as that obtained by injection without the particular risks which that route entails. On the other hand it is less economical since some of the smoke is lost into the atmosphere, and demands better quality heroin. Quite apart from this, some people find that the act of sticking a needle into themselves becomes an essential part of the reward process.

Injectors usually mix the powder with water in a small receptacle such as a spoon, heat the mixture and add a little acid in the form of lemon juice or citric acid to get it to dissolve, then draw it into the syringe through a small piece of cotton wool (or a cigarette filter) to get rid of the larger chunks of undissolved matter. This is a particularly important part of the operation when one is injecting ground-up tablets which contain large amounts of chalky material. When times are hard, junkies will boil up their 'cottons' to extract every last bit of intoxication or comfort from their investment. The next step is to apply a tourniquet ('tie-off'), locate a vein and 'shoot up'. Many people have the habit of drawing blood back into the syringe once or twice ('booting') to ensure that all the material has been absorbed.

There are thousands of first-hand descriptions of the opioid experience in print. Here is one from a nineteenth century physician who was a frequent indulger.

'A sensation of fullness is felt in the head, soon followed by a universal feeling of delicious ease and comfort, with an elevation and expansion of the whole moral and intellectual nature which is, I think, the most characteristic of its effects ... the intellectual faculties are raised to the highest point compatible with individual capacity. It seems to make the individual, for the time, a better and greater man. The hallucinations, the delirious imaginings of alcoholic intoxication are, in general, quite wanting. Along with this emotional and intellectual elevation, there is also increased muscular energy; and the capacity to act, and to bear fatigue, is greatly augmented.'

The modern chaser or injector of heroin is likely to report a powerful rush of pleasure, perhaps amounting to ecstasy, as the drug abruptly washes over the brain. This subsides into a delicious relaxed, dreamy, cocooned feeling as anxieties and fears melt away. The user's head may droop and his eyes close as he goes 'on the nod' or 'gouches out'. The body feels heavy, warm. If problems come into the mind at all, they are suffused with optimism and confidence that all will turn out well. Unless overdosed into unconsciousness the mind remains active, though it may appear otherwise to an observer. Nausea and vomiting may occur, especially in inexperienced users. Diconal has a more 'wiring' stimulant effect. The best high of all is said by some to be produced by a mixture of heroin and cocaine in the same syringe—the famed and potentially lethal 'speedball'.

Many addicts are quite blasé about their experience with heroin, and downplay the ecstasy angle considerably. After a while, they say, you are taking the drug just to 'stay straight'. Others tell you they are either stoned or 'clucking' (withdrawing) with no happy medium. All emotions are suppressed in the regular user who is detached and insulated from all cares and worries. Instead of the usual array of problems that life presents, the addict must face only one—where's the next hit coming from?

When used in a medical context, the opioids are remarkably safe, non-toxic drugs. Unwanted effects are for the most part minor and include nausea and vomiting, reduced appetite, constipation, drowsiness, and apathy. Much more rarely patients may experience dry mouth and sweating, allergic reactions, difficulty in urinating, or problems due to spasm of various tubes leading from the liver to the intestines or the kidneys to the bladder. For the vast majority of patients with serious pain, the beneficial effects vastly outweigh the nuisance of any side-effects that do occur. Recreational users, naturally, may also experience these effects from time to time.

In overdose, of course, the opioids are anything but safe. Depression of breathing can easily lead to a fatal outcome if not recognized and treated, drowsiness may progress to unconsciousness, and the blood pressure may fall so low that vital organs are starved of their oxygen supply. Some individual opioids have particular risks in overdose; for example, pethidine produces a break-down product that can induce fits.

The greatest risks of non-medical use stem from the uncertain strength and purity of street drugs which make overdose or poisoning an ever-present hazard (at the time of writing, at least six people have died within a week in Bristol as a result of a batch of heroin which was 80 per cent pure), and life-style factors consequent upon addiction (see below). The risks are heightened by the pre-existing maladaptive personality characteristics of some users, and by the compulsive, dependent style of drug use that many adopt.

It is very hard indeed to use opioids without becoming addicted to them. Having said that, there are undoubtedly some people who do manage this, but very little is known about them because they mostly remain invisible to services. Those who are coping successfully with their lives do not need the help of doctors or social workers. It is therefore unwise to build a picture of a typical opioid user based on those who stagger through the doors of Drug Dependency Units, or pester their doctors incessantly for prescriptions. This would be as misleading as trying to understand what social drinking was about by talking to a group of skid-row alcoholics.

So what do we know about these non-addicted opioid users (sometimes referred to as 'chippers')? First, we have no real idea how many there might be. It has been suggested that there may be more chippers than addicts, and certainly most dependent users will say they know people who seem to be able to use intermittently. Population surveys demonstrate that far larger numbers have used heroin at least once than ever become addicted or require treatment. It must be the case, therefore, that many people are able to experiment a few times and avoid going on to regular use. What is less certain is whether it is possible to use intermittently for long periods without becoming hooked. Are such intermittent users simply in a transitional state on a path to either giving up or becoming addicted?

Research suggests that intermittent use is determined more by an ability to build the drug into a social ritual than by personality factors or family background. For example, a person may limit heroin use to a particular time and place, and separate it completely from family life and the social circle this includes. On the other hand, common sense dictates that the more heroin (or any other drug) is able to compensate for deficiencies or disadvantages that distress a person, the more difficult it is going to be for that person to relinquish the drug or use it sparingly.

How do chippers differ from addicts? Most retain jobs, homes, and families. They are unlikely to be involved in criminal activities, except a minority who take part in some drug-dealing, usually on a small scale. Most restrict the frequency of heroin use to once or twice a month, with 20 per cent using less frequently and 20 per cent using weekly. At least some of them move into and out of dependent use from time to time; 20 per cent in one series of interviews gave a history of addiction in the past.

Experience gained from heroin use by American servicemen (GIs) in Vietnam is relevant to this discussion. The drug was extremely accessible and very cheap, and large numbers of GIs snorted, smoked, and even injected heroin during their active service. Moreover, with a purity of over 90 per cent (compared with 5–10 per cent on the streets of New York), this was highly potent material. It is interesting to note that despite the easy availability of high quality drugs, most of these individuals never became hooked and did not use heroin on leave in the US, or when their tour of duty finished. Of those that did become addicted in Vietnam, only 20 per cent ever used in the US. Even among those who did go on to use back home, only about 12 per cent were addicted there and only one per cent were still using the drug a year after their return.

Even intermittent users of street heroin expose themselves to considerable risk, however, especially if they choose to inject it. Variations in heroin content may result in accidental overdose and people who have not developed the tolerance that comes from daily use may actually be more vulnerable to this. Poisoning by adulterants and impurities, or life-threatening infections, are always a possibility. The latter may result from bacteria and viruses being carried in the blood and taking hold within the heart, liver, lungs, bones, or brain, as well as a very dangerous type of infection within the blood itself (septicaemia). Local skin infection leading to abscesses is commonplace. Sharing needles or syringes is one of the most important ways in which infectious hepatitis and the human immunodeficiency virus (HIV) which leads to AIDS are passed on.

Even though some people may be able to avoid it, it is difficult to use heroin regularly without becoming dependent. Concepts of addiction are defined and discussed in detail in Chapter 10. Here it is simply enough to note that opioids cause both psychological and physical dependency. The former involves an intense desire for the drug (craving), and a reorganization of one's life around ensuring a regular supply. The latter refers to the physical symptoms which appear when the drug is no longer available. Once addiction is established, people will go to extraordinary lengths to maintain their habit. Consideration of the rights of others or the consequences to their own life and well-being tends to go out of the window.

You have to work at building up a physical dependence on heroin as it is not something which develops after a few fixes. People vary considerably in their vulnerability, but a couple of weeks at least of daily use are likely to be required. The usual pattern is a gradually increasing frequency of intermittent use over many months, with a blithe confidence that despite everybody's warnings, 'I can handle it'. Tolerance to the drug's effect, and the consequent steady increase in dose starts the rot. The early signs of dependence, similar to a particularly stubborn head-cold, are explained away somehow. The realization of addiction then seems to come suddenly as insight breaks through. At this stage the person has to decide whether to seek help, or increasingly embrace the 'junkie' life-style. At any one time in America, there are likely to be around 200 000 people receiving treatment for opioid addiction, and a further 100 000 in prison for opioid-connected offences.

Even when addiction has taken hold, it is clear from historical analysis and some modern case histories that as long as a regular supply of pharmaceutically pure drug is available, and the personality of the addict is adaptive, a productive life-style remains entirely possible. For example, there are a number of documented cases of doctors having long and sometimes highly distinguished medical careers whilst addicted to huge doses of intravenous morphine or heroin. Indeed, Dr William Halsted allegedly asserted that his illustrious career would have been impossible *without* morphine, which he believed increased his stamina and capacity for hard work. As we shall see in Chapter 12, some people argue today that such observations should lead to a reorientation of policy that would lead us back towards the old 'British system' which in the past has been more of a fantasy than a reality.

When someone who has become physically dependent on heroin is denied the drug for one reason or another, a predictable sequence of symptoms called a withdrawal syndrome develops. Roughly eight hours after the last dose, the person is aware of increasing anxiousness, restlessness, and irritability. There is intense desire for another hit or chase. The victim yawns, stretches, sweats, and the eyes and nose begin to stream, with frequent sneezing. Cramping pains in the abdomen gather momentum, nausea comes on in waves and the addict may be embarrassed by uncontrollable diarrhoea. Trembling, deep aching in the bones and muscles, terror and insomnia complete the wretched picture. The skin is pale, clammy and covered in goose-bumps (cold turkey), and the legs twitch and thrash in the bed (kicking the habit). Very rarely, there may be fits or a confusional state with delusions and hallucinations.

These unpleasant but generally not dangerous symptoms are at their height for two to three days then gradually pass off over the next couple of weeks, although sleep disturbance and a feeling of not being quite right sometimes persists for many weeks. The onset of, and recovery from, the syndrome can be delayed if the addict has been taking regular methadone because of this drug's long duration of action. Although the syndrome is certainly unpleasant, most addicts have a disproportionate fear of 'clucking' and their whole life is organized around ensuring at all costs that it doesn't occur. Unfortunately, the exigencies of street life make it inevitable that it will have to be faced on numerous occasions in an average addiction career.

The lifestyle of an addict often involves very poor nutrition and living conditions, and a general lack of self-concern. Women may be driven to exchange sex for drugs, or to take up prostitution to support a habit or keep a family together. It can be very hard for addict parents to give their children the physical and emotional nurturing they need, but many somehow manage to achieve it. Some could benefit from seeking assistance and support from their local drug service, but hesitate to do so for fear that their children will be taken away from them. In fact, this occurs very rarely indeed. These days, doctors and nurses will be concentrating on providing the sort of practical help which will enable these parents to cope more effectively with the problems of family life.

It is not surprising with all these hardships that addicted intravenous drug users have a high incidence of physical and mental health problems. Quite apart from the risks of overdose and poisoning, they are highly susceptible to accidents and serious infection. If a group of addicts is followed up for ten years or so, around 15 per cent can be expected to have died. A study in Rome showed that the death-rate among male addicts was ten times higher than among men of the same age in the general population; for women, the ratio was twenty times.

Opioids have a tendency to suppress ovulation and make pregnancy less likely, but if conception does occur special problems can arise. The addict lifestyle is hardly conducive to a relaxed pregnancy or good ante-natal preparation. Poor nutrition can cause anaemia and increased susceptibility to infection. The strain of coping alone can lead to depression with a risk of self-harm.

There is no evidence that opioids cause direct damage to the fetus, although impurities in street heroin always pose a threat. The main danger is premature delivery of a low birth-weight baby, and there is some increase in the risk of perinatal mortality. It is essential that the mother receives coordinated care and support throughout the pregnancy, and that steps are taken to get her off street drugs whilst avoiding withdrawal symptoms (which may cause miscarriage). This is usually achieved by switching her to methadone, and then either withdrawing this very carefully or maintaining her at the lowest dose consistent with resisting the temptations of the black market. If the birth takes place while the mother is still receiving methadone or taking street opioids, specialist attention from a paediatrician is advisable since the baby is likely to exhibit withdrawal symptoms such as irritability, twitching, diarrhoea, and vomiting, or repeated sneezing during the first few days of life. The onset may sometimes be delayed for days or weeks if the mother was taking methadone. Usually the syndrome responds to simple measures such as gentle handling, demand feeding, and possibly swaddling, but medication (tiny doses of morphine or chlorpromazine) is sometimes required. If the mother has been taking large doses of opioids up to delivery, the baby may not breathe adequately or may develop fits. With appropriate medical management, however, the immediate prospects are excellent. Unfortunately, the lifestyle of an addicted mother may prove hazardous to the physical and mental welfare of the developing child.

Psychiatric symptoms are common among opioid addicts presenting to drug units, but true psychiatric illness is unusual. These symptoms usually disappear of their own accord when the immediate problems related to the addiction are alleviated. True mental illness when it does occur can usually be shown to have predated the addiction.

It is often evident among the people who present to a drug unit that their dependency upon drugs is the least of their problems. This is a highly selected group who characteristically give a history of a deprived, disrupted childhood possibly with physical or sexual abuse, and a record of emotional or behavioural problems which predates their drug use. To attribute their difficulties solely or even mainly to opioid addiction, or to hold them up as a dreadful warning to those who question the merits of the current drug laws would be misguided.

Part III

10
The nature of addiction

Addictus; a citizen of ancient Rome who had built up debts that could not be repaid and was therefore delivered by the courts into slavery under his creditor. That heroin and cocaine are potential slave-masters few would question, but what about tea, or sugar, or gambling?

Any individual's opinion as to the meaning of addiction is bound to be coloured to a large extent by the attitudes of the culture to which he or she belongs. Since these attitudes are by no means consistent across time, nation, or even quite small sub-groups within society, the concept is certain to be a slippery one. Just like our beliefs about the drugs themselves, it is based on value-judgements rather than any serious grip on 'the facts'.

Being 'addicted' is to be caught up in the following sequence: a mounting desire to do something; if this is resisted or prevented, a growing anxiety and preoccupation with the act in question; carrying out the act stills the tension, satisfies the desire and briefly eliminates the need; the cycle then starts all over again. It will immediately be recognized that most biological drives conform to this sequence; eating, drinking, sleeping, having sex. But leaving aside natural functions essential for life, there is a large array of human activities which fulfil the addictive sequence but do not involve the ingestion of any substance. People can get hooked on all sorts of activities from train-spotting to hang-gliding, stamp-collecting to jogging, and suffer as a result. There are people whose work-compulsion leads to as much physical and social devastation as any heroin habit. In many addiction units, a variety of such out-of-control behaviours are managed in the same way as dependence on drugs or alcohol.

Obtaining consensus as to what constitutes a 'drug' is by no means easy. A pharamacologist's definition might be a substance that, when taken into the living organism, modifies one or more of its functions. Many people have difficulty in seeing their enjoyment of alcohol and tobacco in the context of drug use, let alone the gallons of tea and coffee that are consumed every day. Yet in England two hundred years ago, people viewed coffee in a rather similar way to how hashish is viewed today—a substance with a highly significant psychoactive effect. Myths come and go. Caffeine is now disregarded as a bland nonentity whilst cannabis is feared and vigorously suppressed.

Many definitions of addiction have been written, and there are some common strands. There is an ever-growing preoccupation which increasingly gets in the way of the ordinary priorities in life, such as family, work, and leisure pursuits. With this comes a sense of compulsion, a feeling of being obliged to do something that at least in part one would prefer not to do. Routines become ever more focused upon ensuring a ready supply, consumption becomes less dependent upon external cues and less inhibited by concern for consequences. More may be required to achieve the desired effect as tolerance develops.

If addiction is tricky to define, explaining it is a good deal trickier. In order to move away from what is inevitably a rather woolly concept there has been a tendency recently to focus on the two independent elements of physical and psychological dependency. These ideas lend themselves more readily to scientific study, and help to suggest treatment strategies.

Physical dependence is present if a predictable pattern of bodily symptoms and signs, the withdrawal syndrome, develops when the drug is withheld. This phenomenon is characteristic of drugs whose primary effect is to inhibit or depress brain function, and these include alcohol, tranquillizers and sleeping pills, and opioids. Psychological dependency consists of a powerful preoccupation and intense longing for the drug (craving) touched off or modified by elements of the environment (cues) which have become associated in the person's mind with pleasant or unpleasant drug-related experiences. A cue could be a room, a smell, a person, a syringe, almost anything.

Once again, however, we have only got as far as labelling what we observe, and are no nearer at this stage to explaining it. Here we have to accept that there are no established truths, only more-or-less convincing theories. What is clear is that the old 'disease' concept of addiction has to be abandoned. The essence of this proposal was that having had a period of addictive drug use and then achieved abstinence, re-exposure to the drug would inevitably result in immediate reinstatement of compulsive drug use. This simply is not the case. Every clinician knows of a once-dependent drinker who can now manage social drinking, even though this may be a rare beast. Moving in and out of dependency on opiates is more the rule than the exception. Community studies of cocaine use show that the commonest natural history in users who never seek treatment (and would thus still be 'invisible' if they had not been uncovered by field research) is to increase gradually to a peak, then decrease to a much lower level which may be maintained for a considerable period. Currently, the most influential theories fall into three main categories: biological, psychological, and socio-cultural.

Biological models

Biological theories concentrate mainly on the many ways in which drugs can interfere with the transmission of chemical or electrical messages in the brain and nervous system, or alter the balance of neurological function.

The discovery of the endorphins (the 'brain's own opiates') in the 1970s brought home the possibility that drugs may compensate for an in-born or acquired chemical deficiency. These endorphins are chemical messengers made in nervous tissue and released in response to painful, frightening, or satisfying experiences (e.g. sexual intercourse). They turn off aversive (unpleasant) messages and mediate reward. Evidently, it is highly desirable to have an abundant supply at your disposal. If you are deficient in endorphins and happen to be exposed to opiates, or short of dopamine and exposed to amphetamine, it seems logical to suppose that you will be particularly sensitive to the rewarding or aversion-reducing effect. We know from research into depression that people can be naturally low in certain chemical messengers (for example, serotonin), and that this may increase their susceptibility to suicide. Perhaps this deficiency is genetically programmed, an idea which finds some support in animal experiments. This finding has prompted the suggestion that some addicts may be no more to blame for their state than diabetics. Just as a diabetic needs insulin to maintain normal functioning, perhaps the endorphin-deficient junkie needs heroin and the dopamine-starved cocaine fiend must have his stimulant.

Biological systems are generally programmed to restore themselves to their original state when perturbed (known as homeostasis). When the balance is disturbed by an external agent such as a drug, compensatory mechanisms cut in to restore the status quo. These might include activation of feedback mechanisms, alterations in receptor numbers or sensitivity (receptors are the places on cell membranes where chemical messengers of a particular shape can bind) and release of other chemicals which exert an opposing effect to the interloper.

Messages in the brain cross the gaps (synapses) between nerves by means of chemical messengers (neurotransmitters), which are manufactured and stored in the end of the nerve and released by changes in the electrical charge which is actively maintained across the nerve membrane. They pass into the gap, latch on to receptors on the other side, and cause changes in the permeability of the membrane. As a result of this, charged chemicals are allowed to move in or out of the cell, which reconverts the impulse to an electrical wave. There are also receptors on the proximal side of the gap (pre-synaptic receptors) and when the messenger combines with these, further discharge from the nerve is inhibited. Thus the more chemical pours into the gap, the more powerfully will these pre-synaptic receptors inhibit further release. This is called a 'negative feedback loop'. By this sort of mechanism, heroin could induce a long-term shortage in endorphins, and cocaine a persistent impairment of dopamine function. Perhaps it is this secondary deficit state which drives drug-seeking behaviour.

The numbers of the various receptors, and their sensitivity, is not fixed and can be influenced by drugs themselves, or the effect drugs have on naturally occurring brain chemicals. If a drug produces a push in one direction, the brain may increase its output of a chemical which pushes in the opposite direction. The results of this chemical merry-go-round may be manifested in symptoms, behaviours, or emotions.

Withdrawal syndromes may be due to the chemical imbalance left as a drug wears off. For example, one of the effects of opioids is to suppress the activity of the chemical messenger noradrenaline in the brain. If the drugs are taken regularly over a long period, this system becomes distinctly lazy. As the drug wears off and the restraint is removed, it springs back into life with a vengeance. Indeed, the effect is very much like the release of a spring in that there is initially an overshoot. The system becomes hyperactive for a while before settling back to its pre-drug level, and this period of hyperactivity gives rise to many unpleasant symptoms.

Of course, the chemical messenger which is suppressed may not make itself manifest as a physical expression, but a psychological one. Drugs such as cocaine and amphetamine which are not so obviously associated with physical dependency are certainly capable of inducing very powerful mental effects on withdrawal. It seems sensible to assume that these are just as closely related to chemical changes in the brain as the cold turkey of opiate addiction.

But supposing these homeostatic mechanisms failed, either due to the slow recovery rate of suppressed systems or some sort of permanent damage within nerves (as has been found in animals studies with amphetamines, cocaine, and MDMA—see Chapter 7)? An example of the former mechanism may be found in the case of the benzodiazepines. Some people who try to give them up after long periods of continuous use find that they continue to experience distressing symptoms for months, if not years. Some researchers now believe that this may be due to very long-term changes in the receptors for a particular inhibitory chemical messenger, gamma amino butyric acid (GABA). If this were the case, dependency could be explained as a manifestation of the individual's need for the drug simply to maintain a normal physiological balance, just to feel normal. Both this concept, and the inherent deficit model, suggest that the logical response to addiction is long-term replacement therapy.

A fascinating recent advance has been the discovery of a brain pathway which seems to be concerned with the experience of pleasure. Pleasurable activities have been shown to be associated with the revving-up of a particular bundle of dopamine-containing nerves originating deep within the roots of the brain and projecting widely through a blob of cell bodies called *nucleus accumbens* to parts of the brain concerned with emotion, pain interpretation, memory, and reasoning (the ascending ventral-tegmental dopaminergic system, VTDS). If a micro-electrode is implanted into the VTDS of an animal, and the animal is then given the opportunity to stimulate itself electrically by repeatedly pressing a bar, it will choose to do this with tremendous enthusiasm and persistence in preference to other rewarding activities such as eating, mating, or sleeping. Indeed, the animal will press the bar many hundreds of times to achieve a single stimulus and, given free access, will continue to press the bar until exhaustion or even death supervenes. It is very interesting to note that most drugs capable of producing addiction have a powerful stimulatory effect within this pleasure pathway. It has been suggested that craving might be the result of activation of memories of past stimulation of the pathway induced by exposure to environmental cues. One of the most potent drugs in stimulating the pleasure pathway is cocaine, but there is a price to pay. In animals, it seems there is the possibility of permanently damaging the system, so that it remains depleted of its main chemical messenger, dopamine. This might account for the clinical observation that some long-term cocaine users experience a prolonged reduction in their ability to obtain pleasure from life after giving up the drug. Ironic indeed if it turned out that pursuit of the ultimate pleasure led to the destruction of the means of experiencing it.

An interesting phenomenon that can be explained by the existence of a pleasure pathway is called priming. Priming is the phenomenon whereby exposure to a small dose of an addictive substance can spark off a full relapse into addiction in people who have abstained for some time, an observation that formed a major plank in the now-unfashionable disease theory of addiction. Priming can be demonstrated in previously addicted but currently drug-free animals trained to bar-press for reward. A small micro-injection of morphine delivered into the VTDS will immediately reinstate compulsive bar-pressing to obtain cocaine. The theory is that anything which tickles up the pleasure pathway will spark off a 'positive appetitive state' with subsequent drug-seeking behaviours. Note that it does not have to be a dose of the favoured drug; anything which twitches the VTDS will do. An obvious implication concerns the possible role of everyday drugs such as caffeine and nicotine which may be capable of this 'tickling-up' process. Could they be a covert factor in inducing relapse in abstinent heroin or cocaine addicts? Recent work suggests that the tickling-up effect can also be induced by environmental cues associated in the addict's mind with previous drug use.

Of course, the role of this pathway can only offer a partial explanation of addictive behaviour in humans, most of whom do not respond to cocaine in the same way as animals in such experiments. The majority of reasonably well-adjusted humans who snort cocaine on a few occasions do not go on to sacrifice everything to the drug. They are not helpless slaves to their pleasure pathways. The ability to think abstractly, to weigh up cost–benefit analyses, to defer reward, and to place social obligations alongside satiation of basic appetites provides an effective counterbalance for most people against the primitive messages from the pleasure pathway.

What about drugs that are clearly addictive but appear to have little activity in this pleasure system, such as the benzodiazepines (tranquillizers)? There is some evidence of the existence of a separate brain system which could underlie the mechanism for the relief of distress as distinct from the appreciation of reward. It makes sense that drugs which are able to neutralize punishment (depression, fear, pain, anxiety) might potentially be every bit as addictive as those which magnify pleasure.

A final example of a possible role that biology may play concerns the manner in which the body transports and breaks down drugs. It is well known that people differ greatly in their reaction to alcohol. Some seem able to drink until the stuff comes out of their ears and never get a hangover, whilst others get a crashing headache after a half-pint of shandy. One reason why this may occur is the different balance of breakdown products (metabolites) that the liver or other disposal mechanisms may produce. If a person happens to be physiologically programmed to be efficient in producing lots of acetaldehyde (an unpleasantly toxic metabolite of alcohol) but less good at moving it on to the next stage in the breaking-down process, alcohol will not prove very rewarding because small doses will produce unpleasant symptoms. This would make it extremely unlikely that the person would be inclined to progress to excessive drinking and addiction. On the other hand, a more efficient metabolism might necessitate drinking large amounts in order to get as mildly tipsy as one's friends. Since it is known that the enzyme systems that drive these metabolic processes are genetically determined, one might expect vulnerability to run in families, as indeed turns out to be the case (though there are other possible explanations for this observation).

Psychological models

Psychological models can be sub-divided into psychoanalytic, behavioural, cognitive, and personality theories.

Psychoanalytic explanations vary according to the school of thought—Freudian, Jungian, Kleinian, and so forth—to which the theorist belongs. Put very simply, the basic concept is that observed behaviour is the result of an interaction between external events and repressed or unconscious mental processes of which the subject remains unaware, unless and until they are revealed and interpreted by psychoanalysis. Merely tinkering with the surface behaviour, in this case drug dependency, is a waste of time. Even if you succeed in getting rid of it, the internal conflicts remain and will become manifest again in due course either in the form of relapse, or by the appearance of some other neurotic activity (symptom substitution). There are various more specific explanations of addiction. Some analysts believe that addicts are regressing to unfulfilled phases of psychosexual development. Others refer to defects in personality structure such as a poorly developed 'super-ego' or retention of 'infantile thought processing'. There is talk of the search for a form of satiation that was never achieved in early childhood, of oral fixation, or 'anal regression'. The drug user may be seen as 'fundamentally suicidal'. Attempts have been made to explain why particular individuals may be drawn to particular drugs. Heroin users, for example, experience '... a need for the ego to control feelings of rage and aggression, emotions that relate to the anal stage of psychosexual development'. The 'weak ego structure' of such individuals leads them to seek quiet and lonely lives, hence the attraction to narcotics.

There was a time when such ideas underpinned the treatment philosophy of many drug and alcohol units, but this is no longer the case. Abstract hypotheses like these do not lend themselves to scientific testing, and even analysts from within the same theoretical school often come up with completely different interpretations of the same observed phenomena. Psychodynamic treatments of addiction have not proved particularly successful, and they are very time-consuming. The psychodynamic approach can sometimes seem patronizing and judgemental. Since it hinges on the ability of some external 'expert' to understand and interpret the hidden meaning of the various behaviours and feelings, it also has the effect of increasing the addict's sense of helplessness and passivity.

Behavioural models are based upon the various forms of learning theory. Classical conditioning is familiar to most people through the efforts of Pavlov and his dogs. If you expose a dog to food (unconditioned stimulus) it salivates (unconditioned response). If you ring a bell (conditioned stimulus) on a number of occasions at about the same time as you present the food, you will find after a number of exposures that the dog will salivate when you ring the bell (conditioned response) even in the absence of food. By the same process, various elements of a drug user's environment (cues) become linked to highs and lows of the drug experience. Thus a cue which has become a conditioned stimulus for euphoria may spark off positive memories of drug use, leading to the craving for a repeat performance, whilst one which induces the symptoms of withdrawal will lead to drug-seeking behaviour to alleviate the misery.

Many injecting drug users find that they become extremely attached to the process of injecting, which itself becomes highly rewarding to the extent that they will sometimes inject themselves with water or other inert substances between 'live hits'. This phenomenon can be explained on the basis of classical conditioning, which also suggests a specific treatment (cue exposure and response prevention—see Chapter 11).

'Instrumental' or 'operant' conditioning is induced according to the outcome of the behaviour. Activities which have a pleasant outcome, or take away some unpleasant experience like depression or fear, are rewarding or reinforcing and are likely to be repeated. Not at all surprising, you might think, but why should the causes of dependence necessarily be counterintuitive or obscure? Experimental psychologists spend much of their time testing out the tenets of common sense.

In animals, there are now a number of well-established experiments by which one can compare the reinforcing properties of different drugs. For example, a rat or a monkey can be trained to press a bar to obtain a reward, and the number of times it is prepared to do so is a measure of the reinforcing power of that particular reward. If the reward is a small amount of amphetamine, cocaine, or heroin, a monkey will press the bar thousands of times to obtain a single dose. If allowed the choice between food or cocaine, it is likely to choose the cocaine option to the point of death. Alcohol, barbiturates, and benzodiazepines show similar results, but induce much lower rates of bar-pressing. By comparing the number of bar presses with the results obtained from standard rewards such as access to food or water, the findings can be calibrated to the real world.

The way in which the reward is presented is also influential. You might think that a consistent, fixed relationship with the activity would be most reinforcing (i.e. compelling) with each press of the bar predictably delivering a gulp of water or chunk of food, but you would be wrong. This is one of a number of examples where common sense proves misleading. In practice, the most reinforcing pattern is an intermittent, random one. This is what makes slot machines ('one-armed bandits') so amazingly addictive. Not all that surprising, then, to discover that these mechanical pick-pockets were invented by a psychologist. I expect he or she forgot to patent the idea, however.

The third form of conditioning is vicarious learning, through contemplation of what happens to others. This 'modelling' can take place after observing the behaviour of family or friends, or activities portrayed in the media. Analysis of the effect of advertising confirms that compelling images such as the 'Marlboro man' do actually succeed in associating, in this case, the act of smoking with feeling tough, independent, and devil-may-care. Repeatedly seeing the heroes (or anti-heroes) of films puffing away on cigarettes or gulping down quintuple whiskies has a measurably reinforcing effect on viewers of all ages. This form of learning can have a powerful influence in directions which may not have been predicted. A recent government-funded anti-drug propaganda programme (with the title 'Heroin screws you up') backfired because far from repelling adolescents from the dreaded panapathogen, it associated it firmly in their minds with satisfying images of rebellious youth daringly rejecting the uncool, safety-first, and wholesome image represented by their parents and teachers.

Psychologists have models to explain all of the important components of dependence, such as withdrawal, intoxication, and craving. Many of these are very convincing though complicated and beyond the scope of this book (and often the comprehension of its author). A point worthy of note is that psychology plays a vital role even in the context of events which appear at first sight overwhelmingly biological. This can be illustrated by reference to the phenomenon of tolerance. Regular, repeated exposure to many drugs leads to increasing resistance to some or all of their effects. Larger doses are required to achieve the same result, and the individual can eventually tolerate doses that would kill an inexperienced user. Pharmacologists can put forward several explanations for this, pointing to alterations in the way the body is metabolising the drug, changes in the number and sensitivity of receptors, and homeostatic responses resulting in the overactivity of pharmacological systems with opposite effects to those of the drug. A simple experiment illustrates that there is more to this than basic pharmacology. If an animal is exposed to gradually increasing doses of morphine in a completely consistent environment, it will eventually tolerate a huge dose without discomfort, as one would expect. If it is then moved to an unfamiliar place and given exactly the same dose, it will keel over and die of an overdose as if it were a drug-naive animal. It has also been noted that drug-addicted rats which have been withdrawn from opiates become re-hooked much more quickly in the cage associated with the previous addiction. These observations suggest that what the animal has learnt to expect powerfully shapes what actually happens, probably by activating homeostatic pharmacological systems.

Cognitive theories focus on the way people interpret their experiences in life, and account for what has happened to them. Emotions are not seen as the product of surging chemicals or the repressed unconscious, but rather as the logical outcome of particular patterns of thinking. These thoughts are usually so fleeting and familiar that subjects are completely unaware of them, and the profound effect they are having in shaping their view of the world and themselves. Cognitive therapists attempt to uncover these 'automatic' thoughts, then collaborate with the patient to see how true or useful they are.

An example of how this works in practice can be illustrated by the experience of an undergraduate who found that he had frequent, inexplicable bouts of gloominess which he self-medicated with cannabis. Unfortunately, the cannabis was interfering with his ability to study and satisfy his tutors, so he sought help. Asked for the most recent example of these sudden attacks, he described being gripped by one whilst on the way to keep his appointment. Retracing his steps in his mind, he was able to place the onset of the mood at a particular point on the street, then remembered that shortly before this point he had nodded to an acquaintance passing in the opposite direction but had received no acknowledgement. Invited to imagine himself back there and examine the thoughts going through his head at that moment, he eventually came up with the following sequence: 'he's ignoring me; he thinks I'm pathetic; he obviously doesn't like me; who the hell does like me?; I *am* pathetic; nobody likes me; I'm always going to be lonely'. Given this interpretation who wouldn't feel depressed, but does it hold water? Perhaps his acquaintance simply didn't see him. Such thought sequences often stem from deep-seated negative beliefs the individual has about himself. Ways in which the cognitive therapist tries to get to grips with these are described in Chapter 11.

The way people explain what has happened to them demonstrably shapes their expectations for the future, and hence what actually comes about. These explanations or 'attributions' can be categorized in various ways, such as internal or external, stable or unstable, specific or general. Individuals tend to be quite consistent in the way they attribute, and this 'attributional style' has a profound effect on self-concept and confidence in achieving desired goals. Imagine three people in the same class who have just failed a maths exam. One is very miserable, one is apparently unmoved, the third is delighted. How is one to account for such different reactions to the same event? The first person thinks he failed because he is stupid. This attribution is internal, stable (he will still be stupid tomorrow and next year), and general (it has implications beyond the boundaries of the current event). No wonder he is miserable. The second person blames the examiner for setting such poorly-worded, inappropriate questions (external, stable, specific); the result has little bearing on his ability or future. The third believes that he is destined to be a great painter (internal, stable, general) and that maths is a complete waste of his time; this result will finally convince his parents that a career in engineering, though secure, is not for him. Most 'normally' functioning individuals automatically adopt a 'self-serving attributional bias', giving themselves the credit when things are working out well but blaming external agencies (luck, the government, God's will) when they are not.

A self-defeating attributional style can lie at the heart of the helplessness expressed by many dependent people. Telling myself 'I've got a disease, it's not my fault, there's nothing I can do about it' is likely to be a self-fulfilling prophesy in that it gives me licence to sit back and wait for someone else to come along with a magic pill or some other passive cure. The subtle, undeclared advantages of continuing an apparently self-destructive behaviour are sometimes overlooked in the face of disadvantages which seem horrific to the observer but, perhaps contrary to what he says, have little genuine impact upon the subject. How desirable, all things considered, the 'addict' *really* regards a particular outcome, such as abstinence, is not always easy to ascertain for patient or therapist, and will obviously influence the likelihood of reaching that outcome (which may be much more attractive to therapist, employer, and spouse than the 'addict' himself). I have put 'addict' in inverted commas here because some attributionally-minded theorists regard the whole concept of addiction as a myth, preferring to see it as a form of behaviour which can be explained entirely in the context of attributions, learning, and interpersonal relationships.

Let us take as an illustration a journalist whose addiction to heroin has cost him his job and his marriage, and who now presents at a drug unit with a massive abscess on his leg from a dirty injection. He begs to be admitted for detoxification and rehabilitation, and convinces all that at last the time has come when the 'game is no longer worth the candle'. On the fifth day of his admission, when he is through the worst, he suddenly announces that he can't take any more and discharges himself. Within half an hour of leaving the ward, he has scored a gram of smack and is 'on the nod' in a lavatory at the bus depot. The medical team, the wife, the employer scratch their heads, and wonder at the mystery of addiction.

The difficulty here is that all these people are looking at the 'problem' from a completely different perspective to the 'addict'. They think of the pain and fear of illness, the loneliness and squalor of life in a bedsit away from the family home, the waste of a promising career. The 'addict' knows that heroin will kill the pain and take away all thought of illness. Worries about long-term risk are always displaced by immediate reward, be it ever so slight and fleeting—think of the cigarette smoker! Family life carries a ton of responsibility and frustration as well as reward and satisfaction. Being in the office at 8.30 a.m. every morning of the week, meeting a deadline for an article on the merits of the Exchange Rate Mechanism, running on the treadmill. Nobody expects a junkie to be tidy and polite, confident and cheery in the face of hassle or boredom, considerate and unselfish. How easy to let it all blow away, to compress all these obstacles and cares and worries into just one question: where's my next fix coming from? It's only when this last dilemma can no longer be answered, when the energy and organization and deviousness have ebbed away, when the will to duck and dive is finally sapped that the addict throws himself, temporarily, on the mercies of the 'caring' professions.

Two other largely cognitive ideas seem important in explaining dependency: self-efficacy and self-esteem. The former, in simple terms, is the degree to which a person believes that she possesses the skills necessary to achieve a genuinely desired outcome. Many distressed heroin addicts, for example, can accept that giving up would be highly desirable but do not believe that they could tolerate detoxification, manage their emotions without the cocoon of opium, or resist temptation to use again if by some miracle they did manage to pack it in. Tackling such beliefs in treatment is influential to outcome.

Self-esteem can be defined as the sense of contentment and self-acceptance that stems from a person's judgement of her own worth. People arrive at this sense by self-appraisal in a number of areas: perceived competence and ability to persevere in valued tasks; an estimation of worthiness and significance in comparison with others; perceived attractiveness to, and approval by, other people; the ability to like oneself in the absence of support from others, or in the face of active criticism and disapproval; and an idea of the value of existence in general. Many therapists in the drug and alcohol field regard low self-esteem as an important causative or maintaining factor, and essential to target in some way during treatment. If you don't like yourself, why should you care what happens in the future?

The idea that there may be such a thing as an 'addictive personality' carries little weight these days. Apart from the lack of any convincing research evidence that such a consistent entity exists, there was an uncomfortably pejorative feel to many definitions. By and large, personality labels in general do not actually explain anything, they just lead in circles. Bob murdered his mother because he is a psychopath; he is a psychopath because he murdered his mother. Experts love labels, but the trouble is that those labels can be extraordinarily undermining to their recipients.

Then there is the problem of teasing apart cause and effect. When confronted with an addict of ten years' standing who seems to have a maladaptive or unpleasant personality, how can one be sure that this is not the effect of all that drug use rather than its cause? A potentially more fruitful approach is to analyse the ingredients which come together to make up a personality (traits) to see if any tend to make an individual more likely to try drugs or become dependent upon them.

In studies of normally functioning adolescents, the only trait which is consistently associated with an increased likelihood of experimenting with drugs is that which relates to sensation-seeking or risk-taking. Those children who climb highest in a tree, ride their bike the fastest down a hill, or swim the furthest out to sea are also the most likely to try whatever drugs are going. Deviant children, those who are bunking off school and committing petty crime, also have a much higher prevalence of drug use.

Attempts have been made to pigeon-hole drug users according to some trait or motivation. One classification separates them into pleasure seekers, conformists (responding to peer pressure), experimenters, and self-therapists. Intuitively, it is the latter group that will be more vulnerable to dependent use. As remarked elsewhere, the more a drug is able to offer a particular individual, the more it overcomes or camouflages a deficit or weakness, the harder it will be to relinquish. Shyness, low self-esteem, poor self-confidence, boredom, fear, anger, loneliness—it is hard to go back to all that once you have found a way to push it out of sight. The more 'normal' a population exposed to a drug, the less will be the casualty rate in terms of dependent use. In the recent cocaine epidemic in the US, the proportion of individuals who became compulsive users was estimated at between 10 and 15 per cent, a figure similar to that found with alcohol. In the population attending an average drug dependency unit the proportion would be vastly higher. This may be explicable in terms of the higher prevalence of psychological disorder or maladaptive personality traits in this latter group, or in more sociological terms (see below).

Socio-cultural models

The socio-cultural perspective is the third important way of looking at addiction. Here the primary considerations are the effects of a person's family in shaping attitudes and behaviour, the nature of society itself, and the manner in which the individual and his peer group relate to these social conditions.

The family, the circle of friends, and the neighbourhood networks are crucial in both starting and maintaining drug use. It is under the influence of these structures that the individual's attitudes to drugs develop, and the first initiation characteristically takes place with a good deal of social ritual in the heart of the peer group. At the beginning of a drug career, the source of supply is almost invariably from within the immediate social circle.

Drug use is much commoner among groups that are poor and deprived, and people from this sort of background are greatly over-represented amongst those seeking help from the average drug unit. These are people with little prospect of an adequate education or material advance, whose families are disrupted, who are daily exposed to a wide range of deviant or criminal behaviour, who have little to occupy their time, who have no reason to admire the values of the affluent classes glimpsed on the street or through the television tube. It would hardly be surprising if they sought an internal escape from such a bleak reality. Yet we know from observations on American GIs in Vietnam that a simple proportionate model of drug use and addiction, that is to say that *x* individuals exposed to a drug will result in *y* addicts, doesn't work. Here, there were a number of environmental reasons that encouraged the use of heroin: the normal cultural and moral restraints were absent; familiar close interpersonal relationships were disrupted; the surroundings were alien, hostile, and frightening; heroin was cheap and easily accessible; and the immediate peer group regarded heroin use as socially acceptable and even desirable. So a large number of people used heroin, very pure material and in large amounts, and perhaps half of these became dependent. The first point to note is that this is a much lower proportion than would be predicted from studies on a drug unit or inner-city populations. Then the fact that a considerable majority of those who did become dependent stopped using heroin immediately they returned home demonstrates that once all these sociological parameters were reversed, the 'addiction' usually evaporated.

Deviant people may start and continue to use drugs simply because it is consistent with the other illegal or anti-social activities in which they are indulging. They may be more vulnerable to heavy consumption or dependency because their particular peer group does not disapprove, or actually values it. The grungy image of the classical junkie is very attractive to some, perhaps because it gives them that sense of belonging, of having a place, that their upbringing failed to provide. Certainly a history of broken homes, emotional deprivation, or physical and sexual abuse is very common indeed among drug unit attenders.

Within any society, people evolve into groupings whose attitudes and social values differ fundamentally. For example, I suspect that the average law-abiding resident of a Newcastle housing estate would write a very different essay on 'What I think of the Police' than a lawyer or accountant from Tewkesbury. Within the larger groupings, there are any number of subcultures each with their goals, aspirations, and values. Membership of groups has lots of advantages: shared interests, aims, and activities to talk about and pursue together; a sense of belonging, 'us against the world'; a romantic sense of risk, perhaps, of not giving a damn, of pushing the limits all the way. Becoming delinquent is not so much a matter of consciously rejecting one lifestyle but gradually learning an alternative one which seems to deliver better results, particularly in the short-term. Values and judgements learned in these groups may quickly lose their influence once the group itself has been left behind. Many a now-successful journalist, businessman, and even the occasional politician would have frequently taken time out from their study of economics or history of art to slip into a kaftan and pass round a joint to the strains of Purple Haze or the Incredible String Band. Many must have gone all the way down the hippy trail, before returning to the path of righteousness as greater reinforcements beckoned.

Anomie (Greek for 'absence of law') was a word used by the sociologist Émile Durkheim to describe a state in which cohesion within society has been weakened to the point that individuals begin to pursue their own goals with little concern for the 'common good'. At times, this state of affairs can almost seem to become consistent with government policy, and certainly appeared that way in the Britain and North America of the 1980s, the decade of the yuppy and the 'if you've got it, flaunt it' philosophy. Margaret Thatcher went so far as to state baldly 'there is no such thing as society'. Of course, she was merely expressing her enthusiasm for individual enterprise, but what if you happen to belong to a part of society that decidedly has *not* 'got it'? What if you can see how agreeable it would be to have it, and can also see the majority of the population-at-large achieving it in some degree, but no matter how industrious or creative you may be there is no chance at all that *you* are going to get it. Let us imagine an impoverished woman in a ghetto who wishes to become a lawyer or own a Mercedes-Benz, but perceives this is beyond her reach no matter how hard she strives. A person in that position has few alternatives: lower her aspirations, and aim to get a job as a servant travelling to work on a bike; rebel, and become a terrorist or a revolutionary; prostitution; or retreat, perhaps by numbing herself out with drugs. Someone who chooses the last course will have little incentive to stop using drugs even if considerable problems occur as a result.

In summary, the three important theoretical models through which one can set about explaining the observed behaviour which is labelled drug addiction are the biological, the psychological, and the socio-cultural. Whilst it is true that some 'experts' attempt to provide a complete explanation based on one of these models and discount the others entirely, the evidence suggests that they are complementary. All three are likely to play their part in any individual's dependency, though one or two may be dominant for that person. How the models underpin an approach to the treatment of the dependency will be discussed in the next chapter.

Perhaps the most important thing to stress in conclusion is that addiction should not be regarded as a walled-off, standalone 'condition'. Rather, it is a state of mind and pattern of behaviour maintained by a precarious balance of drives stemming from the individual, with his particular genes, biochemistry, and patterns of learned behaviour, and the social and physical environment.

Some people, for internal and/or external reasons, are evidently more vulnerable to becoming addicted than others, but in prospective studies of large groups of adolescents it has proved very difficult to pick out the ones at risk with any accuracy. Once dependency is established, there is no external 'cure'. Addicts have choices and must take responsibility for their actions but, as we have seen, there are immensely powerful physical, psychological, and social forces at work which can make behaviour change very difficult indeed. Treatment can only succeed if it takes place as a collaborative venture, with the addict encouraged to be the prime mover.

11
Helping problem drug users

The first step on the road to obtaining help with a drug-related problem is to recognize that one exists. Not infrequently, the person most concerned remains determinedly unaware that things are getting out of hand, and it falls to friends or family to bring it to his or her attention. This may be the result of 'defence mechanisms', those unconscious mental devices we all use to fend off unwelcome thoughts or inner conflicts, or simply because it isn't the user that has the problem; doing badly at school or acting like a jack-ass at home are, in an immediate sense, problems for teachers and family rather than the miscreant himself.

Of course, it is always possible that these concerned onlookers may be confusing drug use with *problem* drug use. Whilst it is true that any flirtation with an illegal substance has the potential to become problematic in one or more of the physical, psychological, legal, or social spheres, it is far from inevitable that this will come about. Many people are able to keep things under control even while using drugs quite regularly, and don't seem to come to any harm at all.

That being said, regular use of any drug, including tobacco and alcohol, is likely to be a risky and counterproductive activity for adolescents who are still developing physically and emotionally. They are more vulnerable to some of the physical and mental unwanted or toxic effects, and drug use with the rituals that surrounds it can interfere with, or completely disrupt, those activities which open up options for the person in later life. Having options is essential for contentment: perhaps one of the cruellest effects of a drug-fixated lifestyle is that options dwindle away to a single track, with no passing places.

Responding to drug use in the family

There are a number of pointers which may suggest to parents that something is amiss with their child, and that this could be related to drug use. Behaviour around the house may change. An altered pattern of sleeping may become evident, with the child pounding around late at night, then showing a greater-than-usual reluctance to respond to the morning alarm. A previously healthy appetite may fade with resulting weight loss, or alternatively mounds of food are tucked away. A previously placid person may become moody or unpredictable, or a live-wire appear lethargic and torpid. Concentration span may be diminished with restlessness or irritability, and memory may become unreliable. The child may seem suspicious or secretive, hobbies and interests get abandoned, friendship patterns change. He or she may appear unusually pale, tired, and apathetic.

These rather non-specific clues would carry considerably more weight if there were also reports of declining school performance, or the child has actually been seen to be clumsy of demeanour and slurred of speech. The clinching factor would be the discovery of drug paraphernalia around the house, for example peculiar-looking cigarette ends, powder or herbal material in twists of paper, or small plastic bags, tablets, or capsules. Regular solvent inhalers often leave a highly visible trail: odd smells coming from the bedroom or on the breath; plastic bags, cans, bottles, and rags in waste-paper baskets; spill marks on clothes or bedding, faded sleeves from surreptitious sniffing; spots or sores round the nose and mouth, cracked and dry lips, chronic cough and cold.

Discovery of needles, syringes, or ampoules would indicate that problems have progressed into a different league.

How should a parent or guardian react to evidence suggestive of drug use? The most important thing is not to *over*-react, and there are several reasons for this. Firstly, many of the earlier pointers are part and parcel of ordinary adolescence and may have nothing whatsoever to do with drugs. But even if there is some drug use, it is essential not to make matters worse by going off at the deep end and further alienating the child. Quite apart from this, there are some immediate physical risks in winding up the emotional tone in someone who has been using certain drugs recently. A surge of anxiety or fear (or anger) can prove fatal to a child who is high on glue because of the effect on the heart, and could provoke a panic attack or worse in someone deep into an LSD trip.

So first, keep calm. Surveys suggest that a third or more school students have had at least one exposure to illicit drugs before they leave school, and that the vast majority emerge unscathed from the experience. The most constructive approach is to be prepared to listen to what the child says, understand the worries and conflicts he or she is facing, and help to find practical solutions to problems. Gaining the confidence of the child, improving communication, and restoring trust will enable the parent to find out the true extent of the drug use, and develop the sort of collaborative approach that can re-establish more constructive priorities.

To change direction, a person needs to see the prospect of clear benefits resulting from that change. Threats or punishment produce only short-term, superficial responses. Factual information should flow both ways since many teenagers know a lot more about street drugs than their parents or teachers (and doctors). This is why many well-meaning drug-education ventures are ineffective—when some of the content is grossly overstated or at odds with what the recipients see with their own eyes, the credibility of the whole message will be lost. The most important task is to give people the information most relevant to their immediate life-goals, and instigate peer discussion. For example, scare stories about cannabis causing lung cancer or mental illness at some point in the future will probably be discounted by school students—but the fact that it can impair learning or performance in sport is more likely to capture their attention.

Once this collaborative approach is established, the practical steps taken depend on the scale of the problem. There may be health issues which necessitate a check-up from the family doctor. Broadly, the aim is to boost your offspring's confidence and self-esteem by giving a sympathetic ear and practical support, whilst avoiding wading in and taking control. Getting your child to keep a diary can be a useful way of understanding problems, and heightening the awareness of the diary-keeper of the extent of use. Self-monitoring of this sort has been shown to be therapeutic as well as informative in people with out-of-control binge-eating and other counter-productive habits. Various practical ways of solving a particular problem can be discussed but the responsibility for carrying these through should always remain with the young person so that the sense of personal control and self-determination is fostered.

Parents faced with adolescent drug use cannot shirk the need to examine their own conduct and relationships for the source of their child's troubles. It may even be necessary to examine the rest of the family's use of drugs to determine whether the 'do as I say, not as I do' factor might have got a bit out of hand.

Services for problem drug users

For the remainder of this chapter, I will concentrate on the ways which exist for helping individuals with an established drug problem. To start with, what are the basic services which any modern health district ought to be providing?

These should be structured on a number of levels. It is to be hoped that most family doctors will accept that problems related to drugs (and alcohol) fall within their remit, because otherwise the specialist services would be completely swamped. Efforts are being made to improve relevant training to medical students and to include the topic in postgraduate education programmes, and the UK government-sponsored Advisory Council on Misuse of Drugs has issued guidelines for the management of such patients by GPs and other doctors. It is unfortunately the case that the antics of that deviant and criminal minority of addicts whose strategy is to pester or threaten all and sundry in the pursuit of a prescription have given drug users in general a bad name with many doctors, so it is an uphill battle to get across the message that most users are not like this.

It is to be hoped that any big town or city should have some sort of informal walk-in centre where people can get advice, education about HIV (Human Immunodeficiency Virus, which can lead to AIDS), counselling, or referral to relevant services. Outreach work, with street-wise individuals mingling with drug users in their natural habitat giving non-judgemental support and advice, is an essential component. Such workers are often highly effective at spreading the health-protection message to people leading risky life-styles who are not in contact with services, and who have scant regard for the advice of middle-class 'experts' or official pronouncements.

Needles, syringes, and condoms should be made easily available free of charge because of the demonstrable individual and public health benefits that result. Concern about the cost to the taxpayer is completely misplaced; the expenses involved in treating hepatitis and AIDS, not to mention an array of lesser ailments, are in a completely different league. The cost of caring for a single AIDS sufferer for a few weeks may easily exceed that of supporting such services for a whole year.

The worldwide epidemic of HIV has transformed the political and clinical response to injecting drug use, which has been established as a major factor in the spread of the virus. Transmission to heterosexual non-injecting partners and from an infected mother to her baby are of particular concern.

A lack of drug services can be associated with frightening consequences. In Edinburgh, for example, a culture developed among drug users in which the sharing of injecting equipment was routine. The HIV infection rate in a sample of these people in 1987 was a horrifying 52 per cent. Twenty per cent had shared needles with users in other parts of the country. A disaster of this sort has sombre implications for the population at large. In this same city, screening of a sample of people aged between 20 and 30 years attending a particular general practice revealed an HIV positivity rate of one in 14 men and one in 28 women.

Research carried out in needle-exchange clinics suggest that around three-quarters of attenders have no other contact with services, so this represents a unique opportunity to provide health education to an otherwise invisible group. Although knowledge about HIV is good among the majority of attenders, about a third seem to continue sharing needles or syringes, and only a minority use a condom with any regularity. It is still difficult to convince heterosexuals that they too are at risk of acquiring HIV. There is some reassuring evidence that people will change their behaviour in response to health education. Amongst non-attenders, however, outreach work suggests that the majority continue to share equipment from time to time despite, in most cases, awareness of the facts of HIV transmission. There is no evidence that easy availability of needles and syringes results in an increase in injecting drug use, though this remains a theoretical possibility. Public health risks stemming from the careless disposal of used equipment is a very real concern: insisting on exchange of old for new, as far as a chaotic lifestyle will permit, is the best way to counter this risk.

Specialist services are now expected to make every effort to be as accessible and user-friendly as possible. They should be aiming to provide a rapid response, tailored flexibly to the needs of the individual; but in the modern National Health Service with its internal market and tight resourcing it is becoming increasingly difficult to persuade budget-holders to maintain existing levels of funding. This problem is being further aggravated by the steady removal of 'ring-fenced' money for drug services.

On presentation, a patient can expect a careful assessment which will involve detailed history-taking and requests for permission to discuss the situation with a close relative (this would never be done *without* first obtaining this consent—confidentiality is, quite rightly, a major preoccupation). The assessment will include a careful physical examination and whatever further investigations are indicated. Ideally, vaccination will be offered to those who test negative for hepatitis B. HIV-testing, with counselling before and after the test to help people cope and come to terms with whichever result is forthcoming, should be on offer.

Chemical analysis of the urine (or blood) will detect the presence of most street drugs consumed during the last few days or in the case of cannabis, weeks. An important exception is LSD; being so potent that the dose is measured in *micro* grams, there is just not enough to pick up from standard urine assays. A number of laboratory methods are used, including thin layer chromatography (TLC) and more recently enzyme multiplied immunoassay techniques (EMIT). EMIT can detect tiny amounts of drug or metabolite so that even drugs which are destroyed quite rapidly in the body may test positive for several days. More recently it has proved possible to measure the drug content of hair. Like examining the rings on a chopped-down tree trunk, this can give a longitudinal picture of drug use. This interesting and slightly alarming development is likely to prove much too expensive and time-consuming for wide application.

Some people will turn out to have special needs. In the case of a pregnant drug-taker it is most important to give the woman accurate information about the drugs she is taking, pay special attention to her general health and emotional well-being, arrange for support with practical problems, and do all that is possible to promote the most stable possible lifestyle. Certain drugs, such as barbiturates, cocaine, and amphetamine have been linked with direct damage to the fetus. If the mother feels she simply cannot give up, or has a physical dependency, the key task is to wean her away from black-market drugs with a substitute prescription whilst being careful to avoid significant withdrawal symptoms which may bring on premature labour.

The philosophy which now underpins the activities of most Drug Dependency Units (DDUs) is that of 'harm minimization'. In the old days, DDUs were orientated towards abstinence in a fairly rigid way, so that those individuals who wanted help of some sort but were not ready to give up drugs right away got short shrift. These days, driven largely by the threat of an HIV epidemic, there is recognition of the need to make contact with as wide a range of drug users as possible, including those who intend to go on using for the foreseeable future.

Just making contact is not much use in itself if nothing else is achieved, and the minimal requirement in the 'treatment hierarchy' is risk reduction. This involves giving advice about safer sex and injecting, including instructions on how to clean used equipment effectively if fresh 'works' are unavailable. It may be possible to provide basic health care, help with housing, child-care, or legal issues. Many people present with problems partly related to drug use but partly related also to the ubiquitous difficulties of homelessness or 'bedsitterland', lack of money, and dearth of prospects. Such people primarily require practical help, not 'treatment'.

The hierarchical approach should not be taken to imply defeatism. With many individuals it will be possible to substitute oral drug use for injecting, to stabilize the lifestyle to the point that detoxification becomes feasible and desirable, then go on to arrange rehabilitation or strategies to prevent relapse. What this new approach does do is to get things into more sensible proportions. In the words of the Advisory Council on Misuse of Drugs: '... the spread of HIV is a greater danger to individual and public health than drug misuse'. Even leaving HIV aside, saving money in the short term by ignoring chaotic drug users will eventually cost the taxpayer infinitely more in hospital bills, prison and probation costs, street crime, and the break up of families.

One useful way of categorizing drug users, devised initially in connection with cigarette smokers, is to place them into one of four levels according to their current attitude towards their drug. In the case of a smoker, a 'pre-contemplator' is someone happily puffing away at twenty a day without any thought or concern about possible consequences. A 'contemplator' is someone who has begun to think now and again that it would be good to cut down, who experiences the occasional twinge of guilt or anxiety whilst lighting up, or dreads coming across an article about lung cancer or heart disease in the paper (it *could* happen to me!). A person who is actively taking steps towards giving up has reached the stage of 'active change', whilst someone who is now abstinent and struggling to stay that way is at the level of 'relapse prevention'. The point of this classification is to draw attention to the fact that the therapeutic intervention must be appropriate to the stage the person has reached. There is no future in attempting to persuade pre-contemplators to give up; the aim should be to nudge them up to the next stage, to get them to begin the process of weighing up the pros and cons of the habit.

Detoxification

For those who have decided that the time is right to cease drug use, a process of detoxification may be required. This is more often than not a necessity when 'downers' such as alcohol, opiates, or tranquillizers have been taken in sufficient quantity and regularity to induce physical dependence. Alcohol and benzodiazepine detoxification can usually be carried out at home; the former using a diminishing regime of tranquillizers, often Librium or chlormethiazole, with regular supervision from a GP or community nurse, whereas benzodiazepine withdrawal involves slow and steady reduction over weeks or months with some regular counselling throughout. If the individual is dependent upon a short-acting benzodiazepine such as temazepam, withdrawal symptoms are often made more tolerable by transferring to a longer-acting drug, for example diazepam, before beginning the reduction. Reasons for choosing in-patient detoxification from these drugs would include a history of serious complications, such as fits, in previous detoxifications, or co-existing medical or psychiatric problems.

Opiate detoxification is potentially an unpleasant experience feared disproportionately by addicts themselves, but it is not dangerous to those in reasonable health and so can usually be undertaken with confidence out of hospital. Indeed, the average addict will detoxify himself from choice or necessity many times over the months or years of his habituation. Sometimes, all that is needed is brief symptomatic treatment from the GP with diphenoxylate, hyoscine, clonidine or benzodiazepines, or a reducing dose of opiate over a few days.

Unfortunately, detoxification in the form of steady reduction of methadone or some other opiate over days, weeks, or months in the community, or abrupt cessation of opiates with symptomatic treatment of withdrawal symptoms, has a low success rate according to the few published studies and in this writer's personal experience. Perhaps as few as a fifth of attempters successfully resist the temptation to score from the black market. For this reason, addicts are often admitted to hospital for the procedure, which commonly consists of a 21-day methadone withdrawal regime. This has quite a high immediate success rate, though many relapse shortly after discharge. It is important for the patient to be aware from the start that withdrawal symptoms are usually at their height towards the end of the procedure, and may persist at some level for many weeks. Unless the patient is warned of this, it can be most dispiriting to discover that far from feeling invigorated and born-again, the hard-fought-for abstinence leaves him feeling like death warmed up. Small wonder that a call to that old familiar mobile phone number may seem like the best solution. Shortening or lengthening the reduction seems to increase the drop-out rate which, interestingly enough, is not much affected by the size of the starting dose.

An alternative detoxification method is the clonidine-naltrexone technique. This consists of establishing the addict on a regular dose of clonidine, which guards against many of the withdrawal symptoms by inhibiting the 'fight and flight' mechanism through a direct effect in the brain, then gradually introducing increasing doses of naltrexone, a powerful opiate antagonist. This speeds up the withdrawal process, possibly by displacing the opiate from the receptors in the brain. Apart from the advantage of a briefer hospital stay of a week to ten days, quicker if necessary, with resulting financial advantages to hard-pressed services, it has been suggested that persistant withdrawal symptoms may be less prominent. Lofexidine is a more expensive alternative to clonidine, better tolerated by some patients; clonidine sometimes has to be discontinued because of its effect of lowering blood pressure.

Results may be influenced by whether the in-patient stay is in a DDU, psychiatric ward, or general hospital bed, and in the case of the first two, whether it is kept locked with strict restrictions on visitors and the coming and going of patients, or operates as an open unit. The locked ward probably achieves a higher completion rate, but may also be associated with higher levels of subsequent relapse, will not be acceptable to some potential customers, and raises civil rights issues. The open arrangement is more user-friendly but is much more likely to be bedevilled by problems associated with covert use of smuggled drugs.

Compulsive users of drugs not associated with physical dependence and people with pre-existing physical or mental illness may also benefit from a short stay in hospital. Most DDUs do not now segregate alcohol and drug users, recognizing that there are more similarities than differences between them. Detoxification on its own has not been shown to have any influence on longer-term outcome. It is only useful if it forms part of some more general treatment plan.

Substitute prescribing

For those not ready to give up drugs, some form of substitute prescribing may be indicated. The various arguments for and against prescribing will be considered in the next chapter. Here, we will simply note that doctors vary widely in their attitude to prescribing to addicts. Some regard it as colluding with dependency and are not prepared to contemplate it under any circumstances, while others at the opposite extreme argue that heroin and cocaine for injection, smoking, or snorting should be readily available to addicts through drug dependency clinics because prohibition has never worked and merely promotes gangsterism. A consensus would be that availability of substitute prescribing forms an essential part of any service for drug users, but that if employed indiscriminately it may encourage or worsen individual dependency, and significantly swell the black market as happened in the late sixties.

The first step for most patients is to substitute methadone for black-market supplies because this drug can be given once a day by mouth, and hence encourages a more stable lifestyle. A disadvantage is that it is an extremely powerful opioid in its own right, and produces a withdrawal syndrome every bit as severe as that of heroin. It is in no sense a cure, merely an effective palliative. Methadone itself is easy to find on the black market, and can be bought for around £10 per 100 mg.

Tables of equivalence are available to arrive at an appropriate stabilization dose, but a safer method is to titrate against visible withdrawal effects such as sweating, trembling, running eyes and nose, yawning, goose-bumps, and vomiting. On the first day, a lowish dose is given and, after a brief observation period, the patient is asked to return the following morning. If withdrawal symptoms are present, the dose can be increased by a modest increment and the process repeated until withdrawal is no longer evident. Once an appropriate stabilizing dose has been identified, the methadone is either reduced and stopped according to a time-scale agreed between doctor and patient, or continued long-term at the same level.

Methadone maintenance has not been fashionable in Britain lately, although it is currently making a comeback, but is well-established in the US. Studies there show that patients remain in treatment much more reliably than in drug-free programmes, and that there are a number of other practical benefits, including lower levels of drug dealing and general criminality, improved physical and mental health, more stability in family life, and lower rates of unemployment. Arguments against maintenance are that it is a social manipulation rather than a treatment, that it colludes with dependency, and that methadone is harder to come off than heroin. Both sides of the argument have weight, so it becomes a political and moral question: does society wish to limit the practical problems resulting from opiate addiction by this intervention, or not?

Prescribed doses of methadone tend to be much larger in the US than Britain, but this is made possible by a marked difference in practice between the countries. In America, methadone is always consumed at the clinic under supervision, whereas most addicts in the UK collect their prescription from a community pharmacist, often on a weekly or even less frequent basis. Inevitably, some will be sold on to the black market in these circumstances, so a daily pick-up arrangement is desirable unless there are good reasons why this is impractical. In Britain, methadone is usually dispensed for oral use as a 1 mg/ml mixture which most addicts are not interested in injecting because of the unpleasant side-effects it produces and the inconveniently dilute formulation. Exceptions to this rule seem to be becoming less uncommon. There are no apparent toxic effects when methadone is consumed orally over long periods.

A relatively small number of addicts receive injectable prescriptions of methadone ampoules, or pharmaceutical heroin from the hundred or so psychiatrists licensed by the Home Office to provide it. Heroin, dipipanone, and cocaine are the only three drugs which require a special licence to prescribe to addicts; any doctor can prescribe any other drug by any route. This is often used as a short-term measure aimed at building rapport and severing links with the black market, though some patients remain stably maintained on injectables for months or years. Either way, it remains a controversial practice.

The results of a famous clinical trial of the 1970s suggest that when choosing which opioid is most appropriate for substitute prescribing, it is not a matter of which is more 'effective' but a philosophical question as to which outcome is the more culturally acceptable. Intravenous heroin was compared with oral methadone. The patients, who were established opiate addicts adamantly requesting an injectable heroin prescription, were randomly allocated to the two groups and followed up for a year by independent assessors regardless of whether or not they remained in contact with services. Seventy-four per cent of the heroin group remained in treatment for the year of study but, of course, all were continuing to inject. Twelve per cent of this group were regularly selling a proportion of their script into the black market. In the methadone group, 12 per cent broke contact with services immediately on being informed that heroin was not available, and by 12 months only 29 per cent of the original sample remained in contact with services. However, 40 per cent of those not in contact with services had stopped regular use of illegal drugs. Overall, the two groups were similar at the end of the study in their average rates of illicit drug use, time spent in the drug culture, unemployment, criminality, and poor general health, but this equivalence concealed a polarization effect in the methadone group; some did unusually well in terms of illicit drug use, crime and so forth, but a similar proportion fared very badly. So which is the better result? Only a politician can decide!

The prescribing of substitute stimulants such as amphetamine or cocaine is even more controversial, and very few doctors are willing to do it at the time of writing. In my experience it can prove useful for a small number of carefully selected patients who can be helped to move away from destructive amphetamine or cocaine use by judicious, time-limited prescribing of dexamphetamine or milder drugs such as phentermine or diethylpropion.

Psychological techniques

Symptoms of anxiety or depression are common in dependent drug users presenting to health services, but usually clear up spontaneously when the problems related to the addiction are sorted out. A small proportion turn out to have a genuine depressive illness, and these are amenable in the usual way to antidepressant medicines or psychological treatments.

A wide range of psychological techniques are available to help people overcome their dependence, but evidence of efficacy is rather limited. Whilst it is clear that being in regular receipt of these various 'talking treatments' is measurably beneficial, it is difficult to demonstrate convincing differences between them.

The cornerstone approach is counselling. A good counsellor will be an excellent listener, and the sort of person who is able to build up a sense of trust and rapport with a wide range of different clients. Having clarified what the person wishes to achieve, the aim will be to guide him or her towards realistic goals, advise, educate, problem solve, and support. More than anything else, an effective counsellor will build up a confidence-enhancing and supportive relationship, collaborating with the client without compromising his or her autonomy. That counselling is effective is established beyond doubt, and it has also been shown that the extent of the benefit is related to the skills of the individual counsellor. It often seems to this observer that some of the most important of these skills are inherent rather than taught: natural warmth and empathy, patience, confidence and inner strength, communication skills. The sort of concrete benefits which have been recorded include reductions in prescribed and illicit drug use, arrests, and convictions. Clients maintained on methadone do even better if they also receive regular counselling.

There is some uncertainty as to whether more specialized psychological treatments can improve outcome beyond that achieved by counselling. One study suggested that both cognitive-behavioural therapy and 'supportive-expressive psychotherapy' were superior to counselling alone, but this may simply have been related to the time spent with subjects—35 per cent more in the two specialist conditions. Interpersonal psychotherapy showed no advantages over monthly brief contacts in a six-month study of patients on methadone maintenance, though both groups improved.

It seems that these more complicated techniques are only indicated for selected patients with particular problems. An example would be a person with a needle fixation, that is a dependence upon the act of injecting itself induced by classical conditioning (see Chapter 10). Here, a particular behavioural approach, cue exposure and response prevention, may prove effective, at least in the short term. This consists of exposing the patient to a cue such as a loaded syringe without allowing him to inject. The intense arousal and desire to inject gradually subsides over an hour or so, and the patient is kept in contact with the cue until this has occurred. This procedure is repeated at regular intervals, and gradually the patient's response to the cue diminishes (habituates) or disappears completely. Although this approach seems quite promising, it is not yet clear if the effect persists or extends beyond the treatment setting to the real world outside the controlled environment of the laboratory.

Two other techniques have become very fashionable in recent years. 'Motivational interviewing' takes as its starting point an attitude towards the client which is fundamentally different from the traditional, paternalistic view of old. Individuals are expected to take full responsibility for actions and their consequences, encouraged in the belief that they have the power to shape their own destiny, and helped in developing the skills and confidence to achieve personal goals. 'Denial', that great let-out clause for unsuccessful therapy, is seen not as an immovable character trait, but rather as a product of a confronting client/therapist interaction without genuine rapport. The therapist tries in various ways to raise the client's self-esteem and confidence, in the expectation that this improved internal concept will become more and more incompatible with self-damaging behaviour.

Many therapists rely heavily on the principles of cognitive-behavioural therapy (CBT). As mentioned elsewhere, the idea of CBT is to uncover the self-defeating patterns of thinking which, it is argued, lie at the heart of such blights as depression, anxiety, and guilt. Various behavioural experiments are thought up to test out the predictions of therapist and client. For a simple example, let us imagine a person who concedes that his heroin use has become rather more regular recently, but is vigorous in his view that he could give it up at any time he wished. The therapist disagrees. Instead of arguing about what is, after all, a matter of opinion, she suggests a little trial: 'just out of interest, why not see if you can manage without a hit until the day after tomorrow'. The beauty of this is that the outcome should prove useful whichever way it goes. Success would result in a psychological boost to the client, whilst failure might bring home an important insight upon which to build. The therapist will also be interested in analysing with the client the rules by which he governs his life, the 'shoulds, musts, and oughts', attitudes and beliefs, habits and rituals, in order to reflect with him on which are helpful and life-enhancing and which are not.

Avoiding relapse

'Relapse prevention' is all about planning ahead so as to foresee high-risk situations, and understanding how subtle lifestyle decisions may bring these about. An example of the latter would be an abstinent alcoholic deciding to keep a bottle in the house just in case an old drinking companion should call round. Risky situations, such as returning to old haunts or friendships, are either avoided or confronted with rehearsed coping responses. Coping is generally more desirable than avoiding since, if successful, it will lead to an increase in self-confidence and esteem, making future relapse less likely. Of course, the converse is also true, so the choice between avoidance and control is a tactical one.

The therapist will be aware that a small lapse may induce what is known as the 'abstinence violation effect'—'that's it, I've had a hit, all that effort's been a complete waste of time, I'll always be a junkie, might as well go the whole hog'—at which point lapse becomes full-scale relapse. Some even plan a 'controlled lapse' with the client so these feelings can be explored with greater force.

The basic planks of relapse prevention consist of predicting likely high-risk situations and planning in advance how to deal with them, increasing drug-free social contacts and avoiding the drug culture, reinforcing the negative memories of drug use as opposed to the positive, providing continuing support and help with problem-solving, and perhaps urging membership of one of the many self-help groups. Sometimes, drug treatment may have a place. Long-term naltrexone, an opioid antagonist, will protect against impulsive scoring because it blocks the 'high', and certain antidepressants such as desipramine have proved helpful in reducing craving in a proportion of compulsive cocaine users.

What are the commoner causes of relapse? It is hardly surprising that low mood, feeling physically unwell or suffering drawn-out withdrawal symptoms, experiencing powerful craving, falling out with families or partners, and pressure from other drug users feature heavily in most surveys. A less obvious danger is feeling good: 'life's so great I must snort a line to make it perfect!'. Some people come a cropper when they decide to test their personal resolve, confident that enough time has passed to have 'cracked it'.

For many people who have become seriously dependent upon a drug, some sort of post-detoxification rehabilitation is essential for them to remain drug free. This may take the form of continuing help with relapse prevention as outlined above, or a more formal residential or non-residential programme. The latter might consist of long-term counselling, individual or group psychotherapy of one sort or another, family or marital therapy, or more pragmatic approaches like help in moving to another district or finding work.

Self-help groups are made up of people with a common difficulty who meet regularly to provide mutual support, encouragement, and advice. Through regular attendance, people struggling to attain or retain abstinence may find the resolve to do so. Groups also exist for the families and partners of such people. There are likely to be a number of self-help groups of one sort of another in any average health district, and they can generally be contacted through a family doctor or the local walk-in counselling clinic. Many individuals find that self-help groups are very effective in giving a sense of purpose and belonging, and provide a credible abstinence role-model for those who have got used to thinking that this is beyond their reach.

One of the better known drug self-help groups is Narcotics Anonymous (NA). This came into being in the 1950s, basing its philosophy on that of the better-known Alcoholics Anonymous. Members meet very regularly, and anyone who is an ex-user or has a desire to stop taking drugs is welcome to attend. Complete honesty is required, and new attenders may be given an experienced 'sponsor' to whom they can turn for advice or support. At the heart of the meetings is the 'twelve steps philosophy' which requires each member:

1. To admit to powerlessness over his addiction, and concede that life has become unmanageable.

2. To believe that a Power greater than himself could restore him to sanity.

3. To undertake to turn his will and life over to God 'as he understands Him'.

4. To make a 'searching and fearless moral inventory' of himself.

5. To admit to God, himself, and to another human being the exact nature of his wrongs.

6. To be entirely ready to have God remove all these defects of character.

7. To 'humbly ask Him to remove his shortcomings'.

8. To make a list of all the persons he has harmed, and be willing to make amends to them all.

9. To make direct amends to such people whenever possible, except when to do so would injure them or others.

10. To continue to take personal inventory, and when wrong promptly admit it.

11. To seek through prayer and meditation to improve conscious contact with God, 'praying only for knowledge of His will for us and the power to carry that out.'

12. Having had a spiritual awakening as a result of these steps, to try to carry this message to other addicts and practise these principles in all his affairs.

Recognizing that the religious orientation would not appeal to everybody, it is stressed that the way of attaining that spiritual dimension which is an essential component of the process is left entirely to the individual to determine. NA members are accustomed to concentrating on 'one day at a time' rather than trying to envisage a life without drugs stretching ahead interminably. This results in the rewards and satisfaction of success being experienced immediately rather than at some indeterminate point in the future.

Residential rehabilitation usually takes place in units practising the so-called 'Minnesota method', or in therapeutic communities of one sort of another. The Minnesota method is a rather ill-defined individually-tailored package applied over a fixed timespan, usually between four and six weeks. It relies heavily on the twelve-step approach, with a tightly structured timetable which includes education, individual and group therapy sessions, skills training, confidence building, and attention to physical health. Families may be involved in the process. After completing the programme, there is an expectation that most clients would continue to attend NA meetings regularly, or move into a hostel or some other sheltered environment.

Therapeutic communities aim to provide a safe, drug-free environment in which maladaptive ways of coping with life's challenges can be confronted by the peer group, and new ways explored. The organization of such communities varies from place to place, and it makes sense for an interested person to shop around in order to find a regime which seems acceptable. Most programmes last many months, and often involve long periods in which contact with family and friends is discouraged or not allowed. This sort of arrangement may not be appropriate or desirable for someone who retains a degree of social structure, such as an intact family or regular employment. In certain circumstances, the courts are prepared to consider placement in a suitable community as an alternative to a prison sentence.

Many therapeutic communities operate as 'concept houses'. This idea originated in 1959 with the Synanon organization of California. Daily life is highly structured with inmates organized in a very formal hierarchy based on seniority and progress through the system. Staff are generally recovered addicts or, more correctly, 'addicts who are not using at the moment' since this is not seen as a condition which can be cured, only put into remission. The aim is to provide a surrogate family in which a global change in life-view can be nurtured. All behaviour and interactions are closely scrutinized by other residents and subject to peer review through regular 'encounter' groups. These potentially harrowing and confrontational sessions may sometimes go on for several hours. Individuals are brought face to face with their behaviour or personal foibles in a very direct way, and exposed to a particularly vigorous form of peer pressure.

Privileges and responsibilities have to be earned, and a system of rewards and punishment is stringently applied. To get into such a community, a number of hurdles must first be cleared, for example becoming drug-free and getting through a detailed assessment procedure. New admissions are accorded very low status, and are expected to perform a range of menial and unpleasant tasks in the service of the community. Shame, guilt, and public humiliation may be used to enforce conformity. Serious transgression of the rules can result in expulsion and immediate cessation of all access to community members. If a person stays the distance, profound changes for the better are claimed but it is difficult to find concrete evidence of these in the shape of well-planned research studies. Reintegration into the outside world after a long stay in a community of this sort can prove problematic.

A major difficulty for all these forms of rehabilitation is finding the money to pay for them. Provision in health and social service budgets is very limited, and this is partly because of the lack of bullet-proof evidence that they are effective. It is often up to the individual to raise the necessary cash. This problem is likely to become more pronounced as central protection of budgets for drug and alcohol rehabilitation disappears, and local purchasers of health care choose to deploy that provision into other budgets. It seems inevitable that a number of rehabilitation units will fold as a result of this process, further reducing the choice available to these most needy of clients.

12
Drug policy—a need for change?

The signatory nations to the Single Convention on Narcotic Drugs (1961) and the Convention on Psychotropic Substances (1971) are required to 'limit to medical and scientific purposes the cultivation, production, manufacture, export, import, distribution of, trade in, use and possession of [certain] drugs'. The list of drugs to be controlled includes raw opium and coca, opiates and opiate-like drugs, cocaine and cannabis in their various forms, brain stimulants, sedatives and sleeping pills, and hallucinogens. Draconian methods for the enforcement of these controls are justified because of the grave public health, social, and economic risks the drugs are said to pose. A licensing system enables the International Narcotics Control Board to monitor the worldwide trade in licit drugs.

The main policy-making organization for the international control of drugs is the Commission on Narcotic Drugs which has delegates from the member states of the United Nations (UN), along with representation from all those non-UN countries which signed the 1961 Convention. The Commission receives information and recommendations from a variety of interested organizations, and instigates surveys and research. Its role is to devise and monitor strategies aimed at the control of drug abuse and to advise governments on appropriate systems of legal restraint.

Regulation in the UK is through the Misuse of Drugs Act (1971) and Regulations (1985). The former divides drugs into three classes according to their perceived potential for causing harm, and the severity with which offenders are to be dealt. To illustrate these distinctions, a dealer in Class A drugs convicted of a first offence in the Crown Court is confronted with maximum penalties of life imprisonment plus an unlimited fine,

Nothing to declare
CHÂTEAU
MARGAUX
FIGEAC

whereas a Class C dealer faces a maximum of up to two years plus unlimited fine. Class A contains cocaine and coca leaf, the stronger opioids (opium, dextromoramide, dipipanone, fentanyl, diamorphine (heroin), levomethorphan, levomoramide, methadone, morphine, pethidine), the hallucinogens (LSD, psilocin, and related substances), tetrahydrocannabinol and hash oil. Barbiturates and similar compounds including methaqualone, the stronger stimulants (amphetamine, dexamphetamine, methylamphetamine, and methylphenidate), the weaker opioids (codeine, dihydrocodeine, and pentazocine [not strictly an opioid, but a *partial opiate agonist*—see Chapter 9]), and cannabis in herbal and resin form make up Class B. Class C consists of dextropropoxyphene (Distalgesic), the weaker stimulants (diethylpropion, phentermine), the benzodiazepines and some other mild sedatives, pain-killers, and sleeping pills.

The Misuse of Drugs Regulations allocates drugs into five schedules in order to define how they must be stored, prescribed, and documented. Schedule 1 contains substances for which there is, at present, no recognized application in conventional medical practice, such as coca leaf, opium, cannabis, and LSD. Doctors cannot prescribe these under any circumstances, and a special licence from the Home Office is required for anybody who wishes to carry out research on them. This research would, in practice, be limited to animal studies since it seems unlikely in the current legal climate that any research ethics committee would give approval for a study involving administration to humans. Schedule 2 contains those drugs with important medical indications which are also seen as posing very great potential for misuse. These include the opioids, dexamphetamine, and cocaine. Because of the rather daunting list of rules, many doctors are unaware that with the exception of heroin, dipipanone (Diconal), and cocaine, any doctor can prescribe any opiate or stimulant to any addict. A special Home Office licence is required to prescribe the three named drugs to addicts, but any doctor can prescribe even these to non-addicts for medical indications. Schedule 3 is for drugs known to be abused, but to a less worrying degree than those in the previous category. These include barbiturate-like sedatives and sleeping pills and a motley range of slimming tablets and milder pain-killers. Schedule 4 entails only a modest level of control and includes the benzodiazepines, and Schedule 5 is reserved for concoctions which contain tiny amounts of substances from higher up the scale.

Is society the better off for all these restrictions on personal behaviour, or is 'the drug problem' we now confront the product of a delusion? The psychiatrist Thomas Szasz (*Ceremonial chemistry* 1985) has drawn a provocative parallel: 'Formerly, opium was a panacea; now it is the cause and symptom of countless maladies, medical and social, the world over. Formerly, masturbation was the cause and symptom of mental illness; now it is the cure for social inhibition and the practice ground for training in heterosexual athleticism ... the danger of masturbation disappeared when we ceased to believe in it: we then ceased to attribute danger to the practice and to its practitioners; and ceased to call it "self abuse"'. Spinoza held that those who try to restrain personal behaviour by force of law are more likely to arouse vices than reform them. Does an emphasis on external regulation undermine the processes of self-regulation that we all have to fall back on in order to contain our pursuit of pleasurable activities within reasonable limits?

Current drug policy centres on three aims: to reduce the supply of drugs; to decrease the demand for them; and to improve treatment for those who develop drug-related problems.

Reducing supply requires the suppression by force of unauthorized cultivation, manufacture, transport across national frontiers, marketing and personal use of controlled substances. This is the 'war on drugs' and it is an expensive undertaking. The bill for police and customs activity targeted on enforcing the drugs laws in the UK for 1988 was £137 million. The cost in Australia for the same year came to A$258 million. And the amount spent by the US government has been estimated at between US$15 and $20 billion yearly. Is this money well spent?

If the yardstick is the price and availability of the drugs on the street, the answer would appear to be no. Surveys indicate that heroin and cocaine were, if anything, easier to obtain in 1988 than 1978. In 1993, the Chief Investigation Officer for Customs and Excise is on record as saying that 'the drugs problem in Britain is worse than it has been in the past', while the Deputy Assistant Commissioner of Police stated that drugs were now increasingly available. Seizures of synthetic drugs (Ecstasy, LSD, amphetamines) have reached an all-time high at the time of writing. The street price of heroin and cocaine has fallen in real terms over the last decade while average purity has remained constant or increased. Even the isolated successes may turn out to be pyrrhic victories. In Australia, a reduction in the availability of cannabis was mirrored by an increase in availability of amphetamine. Temporary shortages of heroin on American streets have been associated with the development of 'designer' drugs, often of horrifying toxicity.

The figures for worldwide production of illicit drugs are not particularly encouraging either. The International Narcotics Control Strategy Report (US Department of State, 1992) contains the following table (table 2).

Table 2 Worldwide net production of illicit drugs

	1987	1988	1989	1990	1991
total opium (tonnes)	2242	2881	3948	3520	3429
total coca leaf (tonnes)	291 100	293 700	298 070	310 170	337 100
total marijuana (tonnes)	13 693	17 455	36 755	25 600	23 650

There are other snippets of information which suggest that things are not moving in the right direction. Mortality from illicit drug use in Australia increased by 40 per cent between 1981 and 1987. Causes of death included AIDS, accidental overdose consequent upon the variable purity of street drugs, poisonous contaminants, and drug-related violence. The prevalence of injecting drug use in Australia has reached one per cent of the total population. Customs and police officers in the developed world seem to accept that only 10 per cent or so of illicit drugs entering a country are intercepted or seized on the street. The country with the biggest drug problem, the United States, has throughout the twentieth century had the most repressive drug policy. Cause or effect?

Far from being discouraged by such evidence which has convinced many observers that the war on drugs can never be won, the British government remains resolutely committed to the cause. According to Tim Rathbone MP, Chairman of the All Party Drug Misuse Group of the House of Commons, 'Arguments for legalization are born of despair. Government actions and political leadership can tackle the awful problems of drug misuse'. There is no indication that the opposition parties dissent from this view.

Why stop adults from using drugs if they wish to? The libertarian argument, as I interpret it, would proceed along the following lines. 'What I do to my body is my own affair, as long as I don't harm anyone else. I may have risky habits, but it is up to me to decide whether the benefits or pleasures of such habits justify these risks. There is some good evidence that a diet rich in saturated fat carries a very significant risk to health, but there are no suggestions that the intake of eggs should be regulated by law. If I choose to devour an omelette and chips followed by chocolate truffles with lashings of double cream every night, it's my look-out and nobody else's. Surely I should be allowed to decide what I put into my own body? But if I grow a few cannabis plants in my greenhouse for my own personal consumption, I face a possible prison sentence and confiscation of my assets. It all boils down to an arbitrary and conditioned concept of "good" and "bad" pleasures imposed on the population by special interest groups, puritans, or do-gooders.'

It seems reasonable to question the justification for maintaining the current distinctions between the legal and illegal recreational drugs, based as they are upon quirks of history. It cannot be to do with personal risks or social disruption, since nobody could possibly doubt the awesome toxicity of tobacco smoke, or the domestic and public devastation associated with alcohol misuse. It cannot be to do with addictive potential, since nicotine is amongst the most addictive of all drugs—85 per cent of teenagers who smoke a single cigarette will go on to be regular smokers at some point in their lives. Surely it can't be a taboo against intoxication, or loss of personal control? Nothing could be more intimidating or depressing than a pack of lager louts reeling toward you on a night-time city street, urinating against walls, smashing windows, howling obscenities.

Anthropologists have argued that medical and sociological research, with its problem-orientated focus within non-representative pathological samples, exaggerates the risks of drug use. Policy makers should be aware of this perspective when interpreting the advice of those who have to deal with the casualties of drug abuse. Many clinicians and other concerned parties have had little or no first-hand contact with the recreational, non-problematic use of drugs other than alcohol, tobacco, or caffeine. They find it hard to concede that such use could exist, tending to equate 'drugs' with inevitable addiction and deviant behaviour, and forgetting that the illegal nature of drug use drives it underground and creates a whole new deviant life-style. Those using drugs without problems, like social drinkers, remain 'invisible'. Only the abusers and the non-copers surface in their clinics. Drug addicts presenting to the average general practitioner or drug dependency unit are no more representative of drug use in the community than street alcoholics are of ordinary social drinking.

A whole industry has grown up around the treatment and control of drug use with large numbers of 'experts' producing a range of conflicting contributions to the legalization debate, which has really not progressed far beyond the sphere of vested interest. There is an understandable reluctance amongst politicians to instigate change, with a preference for high-profile rhetoric along conventional lines rather than the risk of 'a step into the unknown'. This is to underestimate the parlous state of affairs that now confronts us: the tremendous opportunity that prohibition continues to provide to organized crime, and the viciousness and debauchery that is thus sustained; the enormous financial burden of keeping up the war on drugs; the fact that the physical and social problems created by making drugs illegal may sometimes exceed the primary risks of the drugs themselves.

Let us at this stage summarize the main arguments for and against decriminalizing street drugs. Those in favour would propose that self-control driven by education and informal peer pressure is preferable to coercion by the state, and would bring drug policy into line with society's approach to other potentially unhealthy activities such as excessive drinking, overeating, or boxing. Prohibition has proved counterproductive. It has vastly increased the profitability and sophistication of organized crime, it costs billions of dollars in law enforcement world wide and it has not been conspicuously successful in its primary goal of reducing supply. There are more people in prison in North America than in any other democratic country at any time in history—and more than half are there for drug-related crimes. Price and availability of drugs are very similar in countries with harsh and liberal drug policies. When applied to alcohol, it became evident after repeal that the social damage had greatly outweighed the health benefits. Demand for the more problematic drugs is relatively inelastic so that an increase in price has only a limited effect on consumption and is passed on to the public by an increase in street crime. Addicts will switch drugs rather than stop using if supplies are interrupted. Cannabis is bulky and the demand less inelastic so suppression may be more successful, but this may simply provide an impetus for criminals to market more profitable (and dangerous) drugs with greater vigour. Far from protecting individuals from themselves, the necessity of buying adulterated drugs from unscrupulous gangsters has made drug use infinitely more risky. Many pre-prohibition addicts were able to live full and productive lives, with some famous doctors among them. Problems which are primarily social or medical are being reclassified as forensic.

Relaxing the law would result in a large financial saving which could be channelled into drug treatment and education, and a reduction in the risks associated with drug use which would benefit both individual and public health. Addicts could lead more normal lives with the secondary bonus that the meaning and image of the 'junky', which may seem glamorous and attractive to some, would be eliminated. The viability of criminal organizations would be damaged and one of the most important drives to street crime would be removed. Pressure marketing of drugs in deprived urban areas by criminals would become unprofitable. Anthropological studies and historical reviews suggest that abundant supply does not inevitably result in uncontrolled consumption.

Those opposed to changes in the law argue that this would be a dangerous and irresponsible social experiment which would take us into the unknown. These drugs are taken for no other reason than to induce intoxication, which is disruptive and not acceptable in our society. To relax the law would be to signal that intoxication is admissible, and would reduce the incentive to manage life without resorting to drugs. The psychopharmacology of drugs as powerful as heroin and cocaine is such that it would be quite inappropriate for them to be treated as ordinary consumer commodities. There is simply no way of knowing how many new users there might be, but if alcohol can be taken as a guide, it is apparent that levels of consumption within the population are proportional to availability, and that the prevalence of problem use relates to this general consumption. Cannabis may be less dangerous but all drugs carry risks, and we already have more than enough problems with alcohol and tobacco. A valid part of the duty of the state is to protect individuals from themselves. Even if it could be successfully argued that an adult must be allowed to take responsibility for his or her own actions, this is certainly not the case for children and adolescents and relaxation in the law would inevitably mean an increase in their use of drugs. Legal restraint is imperative to protect the public at large from the consequences and costs (for example to the NHS) of drug-taking. The pursuit of pleasure for pleasure's sake is selfish and encourages people to avoid work and take advantage of the welfare state. Although not completely successful, the war on drugs must have delayed or deterred some potential users. Legalization would unleash entrepreneurial activity which would encourage drug use, and because some restrictions would still be necessary you would still be stuck with a black market.

Is there anything to be learnt from examining the policies of other countries? 'Harm reduction' is a familiar concept in treatment these days (see chapter 11), but in the Netherlands for two decades it has also been the guiding principle in defining how the drug laws should be implemented. Whilst still aiming to minimize supply and remain in step with other European countries, the Dutch have placed less emphasis on punishing personal possession and use. The police cooperate with a declared wish to keep addicts integrated in society rather than force them underground. There is an acknowledgement that the effects of repressive drug policy are often confused in the public eye with the effects of drugs themselves. According to a spokesman from the Ministry of Welfare, Health and Cultural Affairs, the misuse of drugs is seen as '... a matter of social wellbeing and public health rather than as a problem for the police and the courts'.

The British government, to its great credit, has moved a considerable way in this direction since the HIV epidemic surfaced, but a major practical distinction in the Dutch approach is the differentiation of 'drugs presenting unacceptable risks' from 'hemp products'. Although at the time of writing it remains technically illegal to possess and trade in small amounts (up to 30 grams) of cannabis, this law is not enforced, resulting in the *de facto* decriminalization of the drug for personal use. The stated reason for this step was to '... avoid a situation in which consumers of cannabis suffer more damage from the criminal proceedings than from the drug itself'. This initiative goes back to 1976, and does not seem to have been associated with any measurable increase in the prevalence of cannabis use by Dutch nationals. Unfortunately, the fact that the police continue to prosecute those who deal in amounts greater than 30 grams (in other words, the wholesalers) means that the legitimate businessman continues to be excluded, leaving the field open to the criminal entrepreneur. One spin-off, desirable or not depending on your point of view, is that quality control has improved considerably as a result of customer discrimination; home-produced Dutch cannabis is now reckoned by some to be the best in the world.

The next important strand of policy is reducing demand for

drugs on the streets. This relies to a large extent upon programmes of education about drugs and skills training targeted on key groups, such as adolescents. A comprehensive prevention programme for schools would contain a number of ingredients. Firstly, it would be necessary to provide accurate information about the drugs, and immediately we encounter problems. It is often the case that the trainers have less first-hand knowledge than the youngsters they are training. If a distorted view of the dangers of the drugs most familiar to the audience is given, all credibility will be lost. The trouble is that the true facts about a drug like cannabis might not seem deterrent enough so there is a temptation to propagandize, and this may engender scepticism for the whole message. Yet it is not necessary to invent risks to demonstrate convincingly that use of cannabis is likely to be highly counterproductive for people taking exams, learning new skills, playing sport, and so on. It is also essential that this educational package deals with alcohol and tobacco in the same terms as the illegal drugs. Moralizing or scaremongering is not merely useless but may actually increase the likelihood of experimentation.

It is clear from American research that measurable improvement in knowledge is not always parallelled by changes in use patterns. Better outcome is noted when relevant skills training is provided alongside the giving of information, with an emphasis on group discussions and student participation. Help in clarifying personal goals and ethical values, ambition raising, techniques of stress management, enhancing self-esteem and confidence, assertiveness training, rehearsing adaptive coping strategies, talking about ways of resisting peer pressure, and exploring imaginative and appealing alternatives to drug use are all important.

In North America, a number of companies have now introduced workplace policies on drugs (and alcohol) which may involve compulsory urine screening, and this practice is begining to catch on in the UK. It is uncertain at present what effect this will have on demand for drugs. Concerns have been expressed about the issues raised for personal freedom, in that the results of such screening may reflect activity outside work hours which has no bearing on work performance. For example, a single cannabis cigarette can result in a positive urine test for three weeks or more.

It is quite obvious that some of the most important demand reduction measures have nothing to do with education or coercion. Reducing the social disadvantage upon which drug misuse flourishes is surely a fundamental requirement. Supporting deprived families more effectively whilst not further undermining their self-respect and desire to become self-supporting, rehabilitating the decayed inner-city environment, providing something for the youth of our urban estates to do other than sniff glue or trash cars, reintegrating the disenfranchised. If these issues are not faced, the drug problem will continue to grow and grow, war on drugs notwithstanding.

The third strand in policy is to maximize the quality and availability of treatment for drug problems. Harm minimization is now the established principle. This is defined, and a variety of treatment options are described in Chapter 11.

In every health district, there should be easy availability of free needles and syringes on an exchange basis, with outreach workers providing help and support to those otherwise out of contact with services. A user-friendly walk-in agency run by street-wise people in an accessible location is essential, as this will attract individuals who would not dream of going to a doctor or social workers. It is to be hoped that whilst maintaining independence, the staff of this agency will be on good terms with local statutory organizations so that needy clients can quickly be connected with the appropriate professionals.

The statutory services should do all they can to provide a rapid and flexible response tailored to the individual's need. The assessment should always include a physical examination, and all intravenous drug users who test negative for hepatitis B should be offered a course of vaccination. Education about HIV should be a matter of routine along with easy access to testing for those who wish it, with counselling before and after the test and absolute confidentiality. The range of services available in each locality should include in-patient and out-patient detoxification facilities, access to a full prescribing service, community based counselling and nursing services, more specialized psychological treatments when indicated, good communications with other local medical services for particular needs such as antenatal care or pain relief, access to social workers and legal advisers, and cross-referrals to residential rehabilitation units and self-help groups. It is vital that the client/patient feels empowered in whatever contact he or she has with the services. This is not just a question of basic humanitarianism; the only way that you can bring about a long-term change in someone's behaviour is through boosting that person's own motivation and determination.

In many areas these principles are being enacted with some success, but there are certain groups of people who consistently do not get the help they deserve. Women, particularly those trying to care for small children, are often very hesitant about making contact because the set-up can seem intimidating and male-orientated, or they fear that their children will be whisked away into care. People from ethnic minorities also often feel disenfranchised. Users of drugs other than opiates, and younger people generally, often find it difficult to imagine that anything on offer would have any relevance for them. These problems can be surmounted if services are organized with a view to local conditions, based on good advice from people in a position to know what is happening on the street, including field workers, street agency staff, community nurses, and general practitioners.

Over the last ten years or so the British government has channelled considerable resources into building up services for drug users, but encroachment on drug and alcohol budgets has to be resisted constantly now that there is an internal market within the NHS. Budget holders are constantly looking for ways to make savings here so as to create developments there, but economies in this field will inevitably come home to roost eventually. Drug and alcohol-related problems do not just go away if they are ignored, they surface on someone else's budget—social services, general hospitals or GPs, police, or prisons. The aim has to be to provide the flexibility and accessibility that will attract drug users with problems into contact with services and give them what they need to stabilize and enhance their lives, whilst not at the same time compromising the interests of the public at large.

So, carefully targeted and credible education about drugs alongside practical training in life skills, and approachable and responsive treatment services. The present government seems committed to these endeavours alongside its determination 'to contribute to world efforts to control production at source and to tackle trafficking and the crime and money laundering associated with it'. Is this enough, or are there more radical steps which could and should be taken here in Britain?

The starting point, I believe, is to accept that recreational drug-taking by a substantial minority of the population is here to stay. The past is the best guide to the future, and there has never in the history of mankind been a tribe, race, or society which has not resorted to mind-altering drugs. It is difficult to see any reason why modern Britain should buck this trend. Secondly, one must acknowledge that for the large majority of these people, no measurable detrimental effects will result from this, just as one finds in the case of alcohol. Of course, the casualties that do occur with alcohol and other drugs are highly visible and tragic, but so are those associated with a host of other risky yet enjoyable activities. Is taking a drug a valid pleasure? Few would dispute that a glass of wine can be life-enhancing, yet some doctors and others at the end of the nineteenth century were arguing that alcohol was so inherently addictive that individuals simply could not be trusted to make decisions for themselves about whether to drink or not. Ardent prohibitionists discounted the moderation and harmless pleasure of the majority in a vain attempt to protect the vulnerable minority. Those reformed addicts who supported firm external control had found a comforting model of addiction which excused the individual in its grip from any personal responsibility for his state.

It seems to this writer that there are two radical, interlinked steps that, in the longer term, would guarantee the admiration of future generations for any Home Secretary who set the ball rolling. The first is to mastermind a radical and broad-reaching programme aimed at alleviating the social conditions which most obviously foster problem drug use. How can one be surprised that many people will resort compulsively to drugs when confronted with chronic boredom, discomfort, or despair? An injection of capital into housing budgets to prevent families being dumped into bedsitters, more resources to give the inhabitants of the more dilapidated estates education and skills training relevant to modern economic realities, and access to leisure activities other than joyriding and dealing crack. The second step is to raise the cash to pay for this enfranchisement by changing the priorities of the war on drugs.

Arguments for changes in the law do not in the least reflect a lack of concern about the undoubted risks that drugs pose, but rather an awareness that the present policy is simply not doing justice to the monstrous predicament we face. As described above, it cannot be said to have succeeded in its primary aims of reducing the availability and increasing the price of drugs. Criminal domination of drug supply leads to adulterated and contaminated products being sold at vastly inflated prices to the terrible detriment of individual and public health and preserves the most profitable felonious enterprise of all time. Every man, woman, and child feels the wind of this evil in the ever-growing insecurity and violence that pervades all our lives. It is sheer complacency to argue that we are already doing all we can to resist this implacable erosion of the quality of life. Those who look for more constructive policies cannot sensibly be dismissed as defeatist; it is those who glumly insist that there are no alternatives who are more deserving of that label.

Few people would dispute that some drugs carry far more risks than others. Injecting heroin is surely more worrying than smoking cannabis, yet much of our police and customs time seems to be focused on suppressing the latter. In 1993, a total of 55.57 tons of illegal drugs were seized, of which 53.16 tons (96 per cent) consisted of cannabis. Although many police forces have adopted a cautioning policy for first-time cannabis offenders this is unevenly and arbitrarily applied. The ethnic minorities and those with unconventional appearances or lifestyles seem to experience discrimination in this regard. It is still possible to go to prison for five years for possession of a small knob of cannabis for personal use, and about a thousand people get a custodial sentence for this offence in the UK every year. All this attention seems grossly out of proportion to the threat posed by the stuff. Similarly, the everyday experience of young people taking Ecstasy at raves is greatly at odds with the hysterical accounts which appear in the tabloid newspapers. In reality, the main risk stems from being sold phencyclidine or dog-worming tablets packaged as MDMA. The dangers associated with the occasional use of pharmaceutically pure MDMA, though genuine and potentially tragic for a tiny proportion of users, must be kept in perspective.

Cannabis has become quietly endemic in our society, and the experiences of the Dutch with their *de-facto* decriminalization should embolden us to follow their lead. However, it is important that the law is changed rather than merely suspended, and that wholesaling becomes legitimate as well as retailing, otherwise domination of the market will remain in the hands of those with criminal connections. Release, the widely respected London-based drugs and legal advice agency, has called for the introduction of licensed premises such as cafes or clubs which could sell cannabis, Ecstasy, and 'poppers' to members, with strict monitoring of quality control and the normal regulations concerning trading standards and consumer protection. Apart from the considerable savings which would result from abandoning the impossible dream of stamping out cannabis use, there would be a healthy taxation revenue to look forward to. Naturally, there would have to be strict limits imposed on the marketing of these drugs, and all the restrictions on use that are currently applied to alcohol. It would seem sensible alongside this development to curtail further the active promotion of alcohol and tobacco, bearing in mind that the latter is responsible for more deaths than heroin, cocaine, alcohol, road accidents, the HIV virus, murder, and suicide all put together. Funding would have to be made available to set up research aimed at monitoring very carefully the consequences of this change of policy in terms of volume of consumption and incidence of drug-related problems.

An important spin-off would be the possibilities for high-quality medical research which would be opened up, including the investigation through controlled clinical trials of the therapeutic potential of tetrahydrocannabinol. Areas of interest include the easing of nausea and restoration of appetite during treatment for cancer and AIDS, muscle spasm in multiple sclerosis and paraplegia, and relief of chronic pain. A recent survey in the UK suggested that a clear majority of doctors think this drug should be available on prescription. Some people argue that the euphoria-producing effect would be a contra-indication, but others see it as a useful attribute in a drug used to treat the seriously ill. The marked shortcomings in the existing knowledge of the human pharmacology and toxicology of both cannabis and Ecstasy could be remedied, in particular the effects of the former on short-term memory and the evidence of neurotoxicity in the latter.

As far as other drugs are concerned, it is a pity that such debate as has occurred has been polarized between the 'do-nothings' and the 'total-legalizers'. There is surely potential for compromise and a cautious step-wise approach between these two extreme positions. Welcome changes in treatment philosophy have already taken place in this country, but the question of substitute prescribing of opiates to addicts has generated an acrimonious debate extending over several decades. A complete spectrum of views still exists among doctors, ranging from those who argue against any form of substitute prescribing as collusive with dependency or ethically unacceptable, to those who advocate that cocaine and heroin should be readily available to addicts through drug dependency clinics.

My belief is that the overriding need is to wrest control of the drugs market from the criminal. This necessitates a move further along the path of controlled availability of opiates and stimulants from doctors with the involvement and support of other interested parties locally, particularly the police. A drug habit would become much cheaper even if the addict had to bear the cost of the drugs to the NHS, so that a deviant career would no longer be inevitable. Reintegration of the drug user into normal society would restore peer pressure as a potent restraint on behaviour. It is certainly possible that such 'normalization' initiatives might result in increased numbers of people using harder drugs, some of whom are bound to become dependent. Although an addiction to pharmaceutical drugs is much less dangerous to health and welfare than a street drug habit, it is highly undesirable and still carries a number of risks for the individual and his or her social circle. For this reason, such initiatives should be organized as controlled trials in order that the effects can be monitored carefully by independent observers. Until this work has been done, all we have to go on is the rash speculations of the 'do-nothings' and the 'total-legalizers'. Whether some increase in individual risk is a price worth paying for the potential benefits to society of undermining organized crime and reducing the profits of drug-dealing is a matter for public debate. Government-sponsored research of this sort would provide a more rational basis for this debate than presently exists.

The implications of the current total suppression of naturally occurring and milder forms of the main classes of drugs, such as opium and coca, is worthy of some consideration. Historically, these have proved to be containable within the fabric of mainstream society. Cracking down on them has resulted, in many countries, in an exploding use of the infinitely more powerful and destructive synthetic alternatives.

There is a trend for those residential rehabilitation units which provide a valid alternative to prison for those who commit drug offences to be stripped of their support by local NHS and social services budget-holders and go out of business. This is a foolish short-term saving which will prove expensive to us all in the long run. Drug users continue to use drugs in prison and simply learn to be more deviant, while non drug-using inmates are exposed to the temptation to take up the habit to ease their boredom or distress.

There is scant political will for changes in drug policy at present, probably because it does not seem much of a vote-winner. The momentum for change can only come from better informed and wider public discussion of the issues. The changes in policy which are feasible for individual nations acting independently are rather limited because of the possibility of 'drug tourism', although this is likely to be a negligible problem in the case of cannabis and other 'soft' drugs. The debate must clearly be conducted on an international stage.

I recognize that many people will object vigorously to some of the ideas expressed above, and no doubt there are a number of alternative ways forward which would be more effective and which have not occurred to me. One thing of which I am certain is that fresh initiatives in combating the blight of drug dependence and the monstrous criminal parasite which feeds off it are urgently needed.

Selected bibliography and references

Books

Abadinski, H. (1989). *Drug abuse: an introduction*. Nelson-Hall, Chicago.

Burroughs, W. (1966). *Junkie*. The Olympia Press, London.

CIBA Foundation Symposium 166 (1992). *Cocaine: scientific and social dimensions* John Wiley & Sons, Chichester.

Cocteau, J. (1957). *Opium: diary of a cure*. Owen, London.

Crowley, A. (1922). *Diary of a drug fiend*. Collins, London.

Davies, J.B. (1992). *The myth of addiction*. Harwood Academic Publications Gmbh, Switzerland.

Galizio, M. and Maisto, S.A. (eds) (1985). *Determinants of substance abuse: biological, psychological, and environmental factors*. Plenum Press, New York.

Gerstein, D.R. and Green, L.W. (eds) (1993). *Preventing drug abuse: what do we know?* National Academic Press, Washington, DC.

Ghodse, H. (1989). *Drugs and addictive behaviour—a guide to treatment*. Blackwell Oxford.

Goodman, L.S. and Gilman, A. (eds). (1982). *The pharmacological basis of therapeutics*. Macmillan, New York.

Gossop, M. (1987). *Living with drugs*, 2nd edition. Wildwood House Ltd, Aldershot.

Grahame-Smith, D. and Aronson, J. (1992). *Oxford textbook of clinical pharmacology and drug therapy*. Oxford University Press, Oxford.

Grinspoon, L. (1975). *The speed culture*. Harvard University Press, Cambridge, MA.

Grinspoon, L. (1977). *Marihuana reconsidered*, 2nd edition. Harvard University Press, Cambridge, MA.

Grinspoon, L. and Bakalar, J.B. (1979). *Psychedelic drugs reconsidered*. Basic Books, New York.

Hoffer, A. and Osmond, H. (1967). *The hallucinogens*. Academic Press, New York.

Huxley, A. (1959). *The doors of perception* and *Heaven and Hell*. Penguin Books, Harmondsworth.

Kerouac, J. (1976). *On the road*. Penguin Books, Harmondsworth.

Kesey, K. (1960). One flew over the cuckoo's nest [(p[. 80)].

Laurie, P. (1967). *Drugs: medical, psychological, and social facts.* Penguin Books, Harmondsworth.
Leary, T. (1968). *The politics of ecstasy.* G. P. Putnam, New York.
Lowinson, J.H., Ruiz, P., Millman, R.B., and Langrod, J.G. (eds) (1992). *Substance abuse.* Williams & Wilkins, Baltimore.
Miller, W. (ed.) (1985). *The addictive behaviours.* Pergamon, New York.
Orford, J. (1985). *Excessive appetites: a psychological view of addictions.* John Wiley & Sons, Chichester.
Reynolds, J.E.F. (1982). *Martindale's Pharmacopoeia*, The Pharmaceutical Press, London.
Russell, R. (1973). *Bird lives!* Quartet Books, London.
Shapiro, H. (1988). *Waiting for the man: the story of drugs and popular music.* Quartet Books, London.
Stockley, D. (1992). *Drug warning: an illustrated guide for parents, teachers, and employers.* 2nd edition. Optima Books, London.
Szasz, T.S. (1985). *Ceremonial chemistry: the ritual persecution of drugs, addicts and pushers*, revised edition. Learning Publications, Holmes Beach FL.
Szasz, T.S. (1992). *Our right to drugs: the case for a free market.* Praeger, New York. 1992.
Trocchi, A. (1966). *Cain's Book.* Jupiter Books, Calder & Boyars Ltd, London.
Tyler, A. (1987). *Street drugs*, revised edition. New English Library, London.
Weil, A. (1973). *The natural mind.* Jonathan Cape, London.
Wolfe, T. (1969). *The electric kool-aid acid test.* Bantam Books, New York.

Reports

Advisory Committee on Misuse of Drugs (1988). *AIDS and drug misuse; Part 1.* HMSO, London.
Government Statistical Service (1991). *Statistics of drugs seizures and offenders dealt with, United Kingdom, 1991.* Research and Statistics Department, London SW1H 9AT.
Institute for the Study of Drug Dependence (1991). *National audit of drug misuse statistics: drug misuse in Britain.* ISDD, London.
Institute for the Study of Drug Dependence (1992). *National audit of drug misuse statistics: drug misuse in Britain.* ISDD, London. (p. 40).
International Narcotics Control Board (1992). *Report of the International Narcotics Control Board for 1992.* United Nations, Geneva.

New Internationalist (1991). *The crazy war on drugs*, Issue **224**, October 1991.

Rathbone, T. and Stewart-Clarke, J. (1992). *It's my problem as well ... drugs prevention and education.* Published by Sir Jack Stewart-Clarke.

Schools Health Education Unit (1993). *Young people in 1992.* School of Education, Exeter University.

Solowij, N. and Lee, N. (1991). *Survey of Ecstasy (MDMA) users in Sydney.* Research Grant Report Series DAD 91–69, Drug and Alcohol Directorate, NSW Health Department, Australia.

Wootton Committee (1968). *Cannabis: Report by the Advisory Committee on Drug Dependence.* HMSO, London.

Selected scientific papers

Chapter 1; Why use drugs?

Brook, J.S., Whiteman, M. and Gordon, A.S. (1983). Stages of drug use in adolescence: personality, peer and family correlates. *Developmental Psychology*, **19**, 269–77.

Brown, G.L. and Linnoila, M.I. (1990). CSF serotonin metabolite studies in depression, impulsivity and violence. *Journal of Clinical Psychiatry*, **51(Suppl)**, 31–43.

Coopersmith, S. (1967). *The Antecedents of Self Esteem.* W. H. Freeman, San Francisco, CA.

Dembo, R. and Shern, D. (1982). Relative deviance and the process(es) of drug involvement among inner-city youths. *International Journal of the Addictions*, **17**, 1373–99.

Galizio, M., Rosenthal, D. and Stein, F.A. (1983). Sensation seeking, reinforcement, and student drug use. *Addictive Behaviours*, **8**, 243–52.

Gorsuch, R.L. and Butler, M.C. (1976). Initial drug abuse: a review of predisposing social psychological factors. *Psychological Bulletin*, **83**, 120–37.

Jaffe, L.T. and Archer, R.P. (1987). The prediction of drug use among college students from MMPI, MCMI, and sensation-seeking scales. *Journal of Personality Assessment*, **51**, 243–53.

Jessor, R. (1976). Predicting time of onset of marijuana use: a developmental study of High School youth. *Journal of Consulting & Clinical Psychology*, **44**, 125–34.

Kandel, D.B. (1980). Drug and drinking behaviour among youth. *Annual Review Sociology*, **6**, 235–85.

Kandel, D.B., Kessler, R.C. and Margulies, R.Z. (1978). Antecedents of adolescent initiation into stages of drug use: a developmental analysis. *Journal of Youth and Adolescence*, **7**, 13–40.

Malhotra, M.K. (1983). Familial and personal correlates (risk factors) of drug consumption among German youth. *Acta Paedopsychiatrica*, **49**, 199–209.

Measham, F., Newcombe, R. and Parker, H. (1993). The post-heroin generation. *Druglink*, **8(3)**, 16–17.

Measham, F., Newcombe, R., Parker, H. (1994). Alcohol and drug use among North West youth. *Bristol Journal of Sociology*, In Press, (Summer 1994).

Robson, P. (1988). Self Esteem—a psychiatrist's view. *British Journal of Psychiatry*, **153**, 6–15.

Sadava, S.V. and Forsyth, R. (1977). Person-environment interaction and college student drug use: a multivariate longitudinal study. *Genetic Psychology Monographs*, **96**, 211–45.

Smith, G.M. (1974). Teenage drug use: a search for causes and consequences. *Personality and social psychology bulletin*, **1**, 426–29.

Teichman, M., Barnea, Z. and Ravav, G. (1989). Personality and substance use among adolescents: a longitudinal study. *British Journal of Addiction*, **84**, 181–90.

Zuckerman, M. (1979). *Sensation seeking: beyond the optimal level of arousal*. Wiley, New York.

Chapter 2; The Consequences of Drug Use

Blackwell, J.S. (1983). Drifting, controlling and overcoming: opiate users who avoid becoming chronically dependent. *Journal of Drug Issues*, **13**, 219–35.

Bradley, B.P., Phillips, G., Green, L., and Gossop, M. (1989). Circumstances surrounding the initial lapse to opiate use following detoxification *British Journal of Psychiatry*, **154**, 354–59.

Cottrell, D., Childs-Clarke, A. and Ghodse, A.H. (1985). British opiate addicts: an 11-year follow-up. *British Journal of Psychiatry*, **146**, 448–50.

Edwards, J.G. and Goldie, A. (1987). A ten-year follow-up study of Southampton opiate addicts. *British Journal of Psychiatry*, **151**, 679–83.

Gossop, M., Green, L., Phillips, G. and Bradley, B. (1987). What happens to opiate addicts immediately after treatment: a prospective follow up study. *British Medical Journal*, **294**, 1377–80.

Gossop, M., Green, L., Phillips, G. and Bradley, B. (1989). Lapse, relapse and survival among opiate addicts after treatment. *British Journal of Psychiatry*, **154**, 348–53.

Judson, B.A. and Goldstein, A. (1982). Prediction of long term outcome for heroin addicts admitted to a methadone maintenance programme, *Drug & Alcohol Dependence*, **10**, 383–91.

Maddux, J.F. and Desmond, D.P. (1982). Residence relocation inhibits opioid dependence. *Archives of General Psychiatry*, **39**, 1313–17.

McLellan, A.T., Luborsky, L., O'Brien, C.P., Woody, G.E. and Druley, K.A. (1982). Is treatment for substance abuse effective? *Journal of the American Medical Association*, **247**, 1423–28.

Mclellan, A.T., Luborsky, L., Woody, G.E., O'Brien, C.P. and Druley, K.A. (1983). Predicting response to alcohol and drug abuse treatments. *Archives of General Psychiatry*, **40**, 620–25.

Milby, J.B. (1988). Methadone maintenance to abstinence: how many make it? *Journal of Nervous & Mental Diseases*, **176**, 409–22.

Mulleady, G. and Sherr, L. (1989). Lifestyle factors for drug users in relation to risks for HIV. *AIDS Care*, **1**, 45–50.

Robins, L.H., (1979). Addict careers. In *Handbook on Drug Abuse*. R.L. Dupont, A. Goldstein and J. O'Donnell (Eds), 325–36. N.I.D.A., Washington.

Segest, E., Mygind, O. and Bay, H. (1989). The allocation of drug addicts to different types of treatment; an evaluation and a two year follow up. *American Journal of Drug & Alcohol Abuse*. **15**, 41–53.

Simpson, D.D., (1979). The relation of time spent in drug abuse treatment to post-treatment outcome. *American Journal of Psychiatry*, **136**, 1449–53.

Simpson, D.D., Savage, J. and Lloyd, M.R. (1979). Follow up evaluation of treatment of drug abuse during 1969 to 1972. *Archives of General Psychiatry*, **36**, 772–80.

Stimson, G.V., Oppenheimer, E. and Thorley, A. (1978). Seven year follow up of heroin addicts: drug use and outcome. *British Medical Journal*, **1**, 1190–92.

Thorley, A. (1981). Longitudinal studies of drug dependence. In *Drug problems in Britain: a review of ten years*. G. Edwards and C. Busch (Eds). Academic Press, London.

Vaillant, G.E. (1988). What can long-term follow-up teach us about relapse and prevention of relapse in addiction? *British Journal of Addiction*, **83**, 1147–57.

White, N.J. (1989). Prescribing in pregnancy. In *Oxford Textbook of Medicine*. (Eds D.J. Weatherall, J.G.G. Ledingham, and D.A. Warrell (Eds). Oxford University Press.

Zinberg, N.E. and Jacobson, R.C. (1976). The natural history of 'chipping' (controlled use of opiates). *American Journal of Psychiatry*, **133**, 37–40.

Chapter 3; Cannabis

Chait, L.D. and Perry, J.L. (1992). Factors influencing self-administration of, and subjective response to, placebo marijuana. *Behavioural Pharmacology*, **3**, 545–52.

Culver, C.M. and King, F.W. (1974). Neuropsychological assessment of undergraduate marihuana and LSD users. *Archives of General Psychiatry*, **31**, 707–11.

Deahl, M. (1991). Cannabis and memory loss. *British Journal of Addiction*, **86**, 249–52.

Dittrich, A., Battig, K. and Von Zeppelin, I. (1973). Effects of delta 9 THC on memory, attention and subjective state. *Psychopharmacologia*, **33**, 369–76.

Erinoff, L. (1990). Neurobiology of Drug Abuse: Learning and Memory. *NIDA Research Monograph 97*.

Feeney, D.M. (1976). Marihuana use among epileptics. *Journal of the American Medical Association*, **235**, 1105.

Grant, I., Rochford, J., Fleming, T. and Stunkard, A. (1973). A neuropsychological assessment of the effects of moderate marihuana use. *Journal of Nervous & Mental Disease*, **156**, 278–80.

Howlett, A.C., Bidaut-Russel, M., Devane, W.A., Melvin, L.S., Johnson, M.R. and Herkenham, M. (1990). The cannabinoid receptor: biochemical, anatomical and behavioural characterization. *Trends in Neurosciences*, **13**, 421–23.

Le Dain Report. (1972). *Cannabis. A report of the Commission of Inquiry into the non-medical use of drugs.* Information Canada, Ottawa.

Mayor's Committee on Marihuana. (1944). *The marihuana problem in the City of New York.* Jacques Cattell, Landcaster, Pa.

Mendhiratta, S.S., Wig, N.N. and Verma, S.K. (1978). Some psychological correlates of long term heavy cannabis users. *British Journal of Psychiatry*, **132**, 482–86.

Rochford, J., Grant, I. and LaVigne, G. (1977). Medical students and drugs: further neuropsychological and use pattern considerations. *International Journal of the Addictions*, **12**, 1057–65.

Royal College of Psychiatrists. (1987). *Drug scenes: a report on drugs and drug dependence.* Gaskell, London

Schwartz, R.H., Gruenewald, P.J., Klitzner, M. and Ferdio, P. (1989). Short term memory impairment in cannabis dependent adolescents. *American Journal of Diseases of Childhood*, **143**, 1214–19.

Thomas, H. (1993). Psychiatric symptoms in cannabis users. *British Journal of Psychiatry*, **163**, 141–49.

Varma, V.K., Malhotra, A.K., Dang, R., Das, K. and Nehra, R. (1988).

Cannabis and cognitive functions: a prospective study. *Drug & Alcohol Dependence*, **21**, 147–52.
Wert, R.C. and Raulin, M.L. (1986). The chronic cerebral effects of cannabis use: 1. Methodological issues and neurological findings. *International Journal of the Addictions*, **21**, 605–28.

Chapter 4; The Stimulants

Chiarello, R.J. and Cole, J.O. (1987). The use of psychostimulants in general psychiatry. *Archives of General Psychiatry*, **44**, 288–95.
Connell, P.H. (1958). Amphetamine Psychosis. *Maudsley Monograph* No. 5, Oxford University Press.
Gawin, F.H. and Kleber, H.D. (1986). Abstinence symptomatology and psychiatric diagnosis in cocaine abusers. *Archives of General Psychiatry*, **43**, 107–13.
Gawin, F.H. and Ellinwood, E.H. (1988). Cocaine and other stimulants. *New England Journal of Medicine*, **318**, 1173–82.
Gawin, F.H., Kleber, H.D., Byck, R., Rounsaville, B.J., Kosten, T.R., Jatlow, P.I. and Morgan, C. (1989). Desipramine facilitation of initial cocaine abstinence. *Archives of General Psychiatry*, **46**, 117–21.
Gawin, F.H. (1991). Cocaine addiction: psychology and neurophysiology. *Science*, **251**, 1580–86.
Griffin, M.L., Weiss, R.D., Mirin, S.M. and Lange, U. (1989). A comparison of male and female cocaine abusers. *Archives of General Psychiatry*, **46**, 122–5.
Kleber, H.D. (1988). Epidemic cocaine abuse: America's present, Britain's future? *British Journal of Addiction*, **83**, 1359–71.
Musto, D.F. (1973). Sherock Holmes and Sigmund Freud: a study in cocaine. *History of Medicine*, **5**, 16–20.
Musto, D.F. (1991). Opium, cocaine and marijuana in American history. *Scientific American*, July 40–47.
Newcomb, M.D. and Bentler, P.M. (1986). Cocaine use among adolescents: longitudinal associations with social context, psychopathology, and use of other substances. *Addictive Behaviours*, **11**, 263–73.
Strang, J. and Edwards, G. (1989). Cocaine and crack: the drug and the hype are both dangerous. *British Medical Journal*, **299**, 337–38.
Shapiro, H. (1989). Crack: a briefing from the Institute for the Study of Drug Dependence, *Druglink*, **4(5)**, 8–11.
Strang, J. Griffiths, P. and Gossop, M. (1990). Crack and cocaine use in South London drug addicts: 1987–1989. *British Journal of Addiction*, **85**, 193–6.
Swerdlow, N.R., Hauger, R., Irwin, M., Koob, G.F., Britton, K.T. and

Pulvirenti, L. (1991). Endocrine, immune, and neurochemical changes in rats during withdrawal from chronic amphetamine intoxication. *Neuropsychopharmacology*, **5**, 23–31.

Chapter 5; Psychedelics and hallucinogens

Abraham, H.D. and Aldridge, A.M. (1993). Adverse consequences of lysergic acid diethylamide. *Addiction*, **88**, 1327–34.

Abramson, H.A., Jarvik, M.E., Kaufman, M.R., Kronetsky, C., Levine, A. and Wagner, M. (1955). Lysergic acid diethylamide (LSD-25): 1. physiological and perceptual responses. *Journal of Physiology*, **39**, 3–62.

Bowers, M.B. (1977). Psychoses precipitated by psychotomimetic drugs; a follow-up study. *Archives of General Psychiatry*, **34**, 832–5.

Budd, R.D. and Lindstrom, D.M. (1982). Characteristics of victims of PCP-related deaths in Los Angeles County. *Journal of Toxicology and Clinical Toxicology*, **19**, 997–1004.

Eisner, B.G. and Cohen, S. (1958). Psychotherapy with lysergic acid diethylamide. *Journal of Nervous and Mental Diseases*, **127**, 528–39.

Grinspoon, L. (1979). 1. The major psychedelic drugs: sources and effects. 2. Psychedelic drugs in the twentieth century. In *Psychedelic drugs reconsidered*, Basic Books, New York.

Javitt, D.C. and Zukin, S.R. (1991). Recent advances in the phencyclidine model of schizophrenia. *American Journal of Psychiatry*, **148**, 1301–8.

Isbell, H., Belleville, R.E., Fraser, H.F., Wikler, A. and Logan, C.R. (1956). Studies on Lysergic Acid Diethylamide (LSD-25). *Archives of Neurology and Psychiatry*, **76**, 468–78.

Matefy, R.E. (1978). Psychedelic drug flashbacks. *Addictive Behaviour*, **3**, 165–78.

Milhorn, H.T. (1991). Diagnosis and management of phencyclidine intoxication. *American Family Physician*, **43**, 1293–1302.

Pierce, P.A. and Peroutka, S.J. (1990). Antagonist properties of d-LSD at 5-hydroxytryptamine$_2$ receptors. *Neuropsychopharmacology*, **3**, 503–8.

Sanders-Bush, E., Burris, K.D. and Knoth, K. (1988). Lysergic acid diethylamide and 2,5-dimethoxy-4-methylamphetamine are partial agonists at serotonin receptors linked to phosphoinositide hydrolysis. *Journal of Pharmacology & Experimental Therapeutics*, **246**, 924–28.

Yesavage, J.A. and Freeman, A.M. (1978). Acute phencyclidine (PCP) intoxication: psychopathology and prognosis. *Journal of Clinical Psychiatry*, **44**, 664–5.

Weil, A.T. (1977). Observations on consciousness alteration: some notes on datura. *Journal of Psychedelic Drugs*, **9**, 165–9.

Chapter 6; The Inhalants

Anderson, H.R. (1990). Increase in deaths from deliberate inhalation of fuel gases and pressurised aerosols. *British Medical Journal*, **301**, 41.

Anon, (Editorial). (1990). Solvent abuse; little progress after 20 years. *British Medical Journal*, **300**, 135–6.

Chadwick, O., Anderson, R., Bland, M. and Ramsey, J. (1989). Neuropsychological consequences of volatile substance abuse: a population based study of secondary school pupils. *British Medical Journal*, **298**, 1679–84.

Cooke, B.R.B., Evans, D.A. and Farrow, S.C. (1988). Solvent misuse in secondary school children; a prevalence study. *Community Medicine*, **10**, 8–13.

Davies, B., Thorley, A. and O'Connor, D. (1985). Progression of addiction careers in young adult solvent misusers. *British Medical Journal*, **290**, 109–10.

Davy, J. (1839). *The collected works of Sir Humphry Davy, Bart, vol III: researches, chiefly concerning nitrous oxide*. Smith, Elder and Co., London.

Ives, R. (1990). Helping the sniffers. *Druglink*, **5(5)**, 10–12.

Ives, R. (1990). The fad that refuses to fade. *Druglink*, **5(5)**, 12–13.

King, M.D., Day, R.E., Oliver, J.S., Lush, M. and Watson, J.M. (1981). Solvent encephalopathy. *British Medical Journal*, **283**, 663–5.

Nicholi, A.M. (1983). The inhalants; an overview. *Psychosomatics*, **24**, 914–21.

Lynn, E.J., Walter, R.G., Harris, L.A., Dendy, R. and James, M. (1972). Nitrous oxide; it's a gas. *Journal of Psychedelic Drugs*, **5**, 1–7.

Ramsey, J., Bloor, K. and Anderson, R. (1990). Dangerous games: UK solvent deaths 1983–1988. *Druglink*, **5(5)**, 8–9.

Richardson, H. (1989). Volatile substance abuse; evaluation and treatment. *Human Toxicology*, **8**, 319–22.

Ron, M.A. (1986). Volatile substance abuse: a review of possible long term neurological, intellectual and psychiatric sequelae. *British Journal of Psychiatry*, **148**, 235–46.

Schwartz, R.H. and Peary, P. (1986). Abuse of isobutyl nitrite inhalation (Rush) by adolescents. *Clinical Pediatrics*, **25**, 308–10.

Westermeyer, J. (1987). The psychiatrist and solvent inhalant abuse; recognition, assessment, and treatment. *American Journal of Psychiatry*. **144**, 903–907.

Chapter 7; Ecstasy

Barnes, D.M. (1988). New data intensify the agony over ecstasy. *Science*, **239**, 864–6.

Chadwick, I.S., Linsey, A., Freemont, A.J., Doran, B. and Curry, P.D. (1991). Ecstasy, a fatality associated with coagulopathy and hyperthermia. *Journal of the Royal Society of Medicine*, **84**, 371.

Creighton, F.J., Black, D.L. and Hyde, C.E. (1991). Ecstasy psychosis and flashbacks. *British Journal of Psychiatry*, **159**, 713–15.

Dorn, N., Murji, K. and South, N. (1991). Abby, the ecstasy dealer. *Druglink*, **6(6)**, 14–15.

Greer, B., Tolbert, R. (1986). Subjective reports of the effects of MDMA in a clinical setting. *Journal of Psychoactive Drugs*, **18**, 319–27.

Hayner, G.N. and Mckinney, H. (1986). MDMA: The darker side of Ecstasy. *Journal of Psychoactive Drugs*, **18**, 341–47.

Henry, J.A. (1992). Toxicity and deaths from Ecstasy. *Lancet*, **340**, 384–7.

Mcquire, P. and Fahy, T. (1991). Chronic paranoid psychosis after misuse of MDMA (ecstasy). *British Medical Journal*, **302**, 697.

Molliver, M.E., Berger, U.V., Mamounas, L.A., Molliver, D.C., O'Hearn, E. and Wilson, M.A. (1990). Neurotoxicity of MDMA and related compounds: anatomic studies. *Annals of the New York Academy of Sciences*, **600**, 640–64.

Pearson, G., Ditton, J., Newcombe, R. and Gilman, M. (1991). An introduction to Ecstasy use by young people in Britain. *Druglink*, **6(6)**, 10–11.

Price, L.H., Ricaurte, G.A., Krystal, J.H. and Heninger, G.R. (1989). Neurendocrine and mood responses to intravenous L-tryptophan in MDMA users. *Archives of General Psychiatry*, **46**, 20–2.

Shapiro, H. (1989). The speed trip: MDMA in perspective. *Druglink*, **4(2)**, 14–15.

Solowij, N., Hall, W. and Lee, N. (1992). Recreational MDMA use in Sydney: a profile of Ecstasy users and their experiences with the drug. *British Journal of Addiction*, **87**, 1161–72.

Whitaker-Azmitia, P.M. and Aronson, T.A. (1989). Ecstasy induced panic. *American Journal of Psychiatry*, **146**, 119.

Chapter 8; Tranquillisers and sleeping pills

Busto, U.E. and Sellers, E.M. (1991). Anxiolytics and sedative/ hypnotics dependence. *British Journal of Addiction*, **86**, 1647–52.

Higgit, A.C., Lader, M.H., Fonagy, P. (1985). Clinical management of benzodiazepine dependence. *British Medical Journal*, **291**, 688–90.

Klee, H., Faugier, J., Hayes, C., Boulton, T. and Morris, J. (1990). AIDS-related risk behaviour, polydrug use and temazepam. *British Journal of Addiction*, **85**, 1125–32.

Lader, M. and File, S. (1987). The biological basis of benzodiazepine dependence. *Psychogical Medicine*, **17**, 539–47.

Lader, M. and Morton, S. (1991). Benzodiazepine problems. *British Journal of Addiction*, **86**, 823–8.

Leonard, B.E. (1989). Are all benzodiazepines the same? An assessment of the similarities and differences in pharmacological properties. *Psychiatry in Practice*, (Special Issue), April, 9–12.

Livingston, M.G. (1991). Benzodiazepine dependence: avoidance, detection and management. *Prescribers Journal*, **31**, 149–56.

Lynch, S. and Priest, R. (1992). Insomnia: who suffers, when and why? *Prescriber*, 5th March, 37–44.

Nutt, D. (1991). Actions of classic and alternative anxiolytics. *Prescriber*, 19th May, 31–4.

Perera, K.M.H., Tulley, M. and Jenner, F.A. (1987). The use of benzodiazepines among drug addics. *British Journal of Addiction*, **82**, 511–15.

Petersson, H. and Lader, M. (1984). Dependence on tranquillisers. *Maudsley Monograph (28)*, Oxford University Press.

Rickels, K., Schweizer, E., Case, W.G. and Greenblatt, D.J. (1990). Long term therapeutic use of benzodiazepines: Effects of abrupt discontinuation. *Archives of General Psychiatry*, **47**, 899–907.

Schweizer, E., Rickels, K., Case, G. and Greenblatt, D.J. (1990). Long term therapeutic use of benzodiazepines: effect of gradual taper. *Archives of General Psychiatry*, **47**, 908–15.

Sievewright, N., Donmall, M. and Daly, C. (1993). Benzodiazepines in the illicit drugs scene: the UK picture and some treatment dilemmas. *International Journal of Drug Policy*, **4**, 42–8.

Warneke, L.B. (1991). Benzodiazepines: abuse and new use. *Canadian Journal of Psychiatry*, **36**, 194–204.

Chapter 9; Heroin and the opiates

Ball, J.C., Lange, W.R., Myers, C.P. and Friedman, S.R. (1988). Reducing the risk of AIDS through methadone maintenance treatment. *Journal of Health and Social Behaviour*, **29**, 214–26.

Bennett, T., and Wright, R. (1986). The impact of prescribing on the crimes of opioid users. *British Journal of Addiction*, **81**, 265–73.

Blackwell, J.S. (1983). Drifting, controlling, and overcoming: opiate users who avoid becoming chronically dependent. *Journal of Drug Issues*, **13**, 219–35.

Caviston, P. (1987). Pregnancy and opiate addiction. *British Medical Journal*, **395**, 285.

Kramer, J.C. (1980). The opiates; two centuries of scientific study. *Journal of Psychedelic Drugs*, **12**, 89–103.

Levinthal, C.F. (1985). Milk of paradise, milk of hell; the history of ideas about opium. *Perspectives in Biology and Medicine*, **28**, 561–77.

Musto, D.F. (1991). Opium, cocaine and marijuana in American history. *Scientific American*, July, 40–7.

Porter, J., Jick, H. (1980). Addiction rare in patients treated with narcotics. *New England Journal of Medicine*, **302**, 123.

Rounsaville, B.J., Weissman, M.M., Crits-Christoph, K., Wilber, C. and Kleber, H. (1982). Diagnosis and symptoms of depression in opiate addicts. *Archives of General Psychiatry*, **39**, 151–6.

Sternbach, G.I. and Varon, J. (1992). Designer drugs. *Postgraduate Medicine*, **91**, 169–76.

Skidmore, C.A., Robertson, J.R. and Roberts, J.J.K. (1989). Changes in HIV risk taking behaviour in intravenous drug users: a second follow-up. *British Journal of Addiction*, **84**, 695–6.

Swift, W., Williams, G., Neill, O. and Grenyer, B. (1990). The prevalence of minor psychopathology in opioid users seeking treatment. *British Journal of Addiction*, **85**, 629–34.

Thomson, J.W. (1984). Opioid peptides. *British Medical Journal*, **288**, 259–60.

Wood, P.L. (1982). Multiple opiate receptors: support for unique mu, delta and kappa sites. *Neuropharmacology*, **21**, 487–97.

Zinberg, N.E. and Jacobson, R.C. (1976). The natural history of 'chipping' (controlled use of opiates). *American Journal of Psychiatry*, **133**, 37–40.

Zinberg, N.E. (1979). Non-addictive opiate use. In *Handbook on Drug Abuse*. R.L. Dupont, A. Goldstein, and J. O'Donnell, (Eds), 303–14. N.I.D.A., Washington.

Chapter 10; The Nature of Addiction

Babor, T.F., Cooney, N.L. and Lauerman, R.J. (1987). The dependence syndrome concept as a psychological theory of relapse behaviour: an empirical evaluation of alcoholic and opiate addicts. *British Journal of Addiction*, **82**, 393–405.

Dimond, S.J. (1980). Consciousness. In *Neuropsychology; a textbook of systems and psychological functions of the human brain*. S. J. Dimond (ed). Butterworths.

Childress, A.R., McLellan, A.T. and O'Brien, C.P. (1986). Conditioned responses in a methadone population; a comparison of laboratory, clinic, and natural settings. *Journal of Substance Abuse Treatment*, **3**, 173–79.

Gossop, M. (1976). Drug dependence and self esteem. *International Journal of the Addictions*, **11**, 741–53.

Lart, R. (1992). Changing images of the addict and addiction. *International Journal of Drug Policy*, **3**, 118–25.

Legarda, J.J., Bradley, B.P. and Sartory, G. (1987). Subjective and psychophysiological effects of drug related cues in drug users. *Journal of Psychophysiology*, **4**, 393–400.

McAuliffe, W.E., Rohman, M., Felman, B., Launer, E.K. (1985). The role of euphoric effects in the opiate addictions of heroin addicts, medical patients and impaired health professionals. *Journal of Drug Issues*, Spring Edition, 203–24.

Room, R. (1989). Drugs, consciousness and self-control: popular and medical conceptions. *International Review of Psychiatry*, **1**, 63–70.

Saunders, B., Allsop, S. (1987). Relapse: a psychological perspective *British Journal of Addiction*, **82**, 417–429.

Stewart, J., de Wit, H. and Eikelboom, R. (1984). Role of unconditioned and conditioned drug effects in the self-administration of opiates and stimulants. *Psychological Review*, **91**, 251–68.

West, R. (1989). The psychological basis of addiction. *International Review of Psychiatry*, **1**, 71–80.

Wikler, A. and Pescor, F.T. (1967). Classical conditioning of a morphine abstinence phenomenon, reinforcement of opioid-drinking behaviour and 'relapse' in morphine addicted rats. *Psychopharmacologia*, **10**, 255–84.

Wilson, G.T. (1987). Cognitive processes in addiction. *British Journal of Addiction*, **82**, 343–53.

Wise, R.A. (1988). The neurobiology of craving: implications for the understanding and treatment of addiction. *Journal of Abnormal Psychology*, **97**, 118–32.

Chapter 11; Helping problem drug users

Bennett, T. and Wright, R. (1986). The impact of prescribing on the crimes of opioid users. *British Journal of Addiction*, **81**, 265–73.

Bradley, B.P., Phillips, G., Green, L. and Gossop, M. (1989). Circumstances surrounding the initial lapse to opiate use following detoxification. *British Journal of Psychiatry*, **154**, 354–59.

Brettle, R.P., Bisset, K., Burns, S., Davidson, J., Davidson, S.J., Gray, J.M.N., Inglis, J.M., Lees, J.S. and Mok, J. (1987). Human immunodeficiency virus and drug misuse: the Edinburgh experience. *British Medical Journal*, **295**, 421–24.

Charney, D.S., Heringer, G.R. and Kleber, H.D. (1986). Combined use of clonidine and naltrexone as a rapid, safe and effective treatment of abrupt withdrawal from methadone. *American Journal of Psychiatry*, **143**, 831–7.

Gossop, M. (1978). A review of the evidence for methadone maintenance as a treatment for narcotic addiction. *Lancet*, **1**, 812–5.

Gossop, M., Green, L., Phillips, G. and Bradley, B. (1989). Lapse, relapse and survival among opiate addicts after treatment—a prospective follow-up study. *British Journal of Psychiatry*, **154**, 348–53.

Gossop, M., Griffiths, P., Bradley, B. and Strang, J. (1989). Opiate withdrawal symptoms in responses to 10-day and 21-day methadone withdrawal programmes. *British Journal of Psychiatry*, **154**, 360–3.

Hartnoll, L., Mitcheson, M.C., Battersby, A., Brown, G., Ellis, M., Fleming, P. and Hedley, N. (1980). Evaluation of heroin maintenance in controlled trial. *Archives of General Psychiatry*, **37**, 877–84.

Joe, G.W. and Simpson, D.D. (1975). Retention in treatment of drug abusers: 1971–72 DARP admissions. *American Journal of Drug and Alcohol Abuse*, **2**, 63–71.

Judson, B.A. and Goldstein, A. (1982). Prediction of long-term outcome of heroin addicts admitted to a methadone maintenance program. *Drug and Alcohol Dependence*, **10**, 383–91.

Kleber, H.D. (1989). Treatment of drug dependence: what works. *International Review of Psychiatry*, **1**, 81–100.

Kreek, M.J. (1979). Methadone in treatment–: physiological and pharmacological issues. In *Handbook on drug abuse*, R.L. Dupont, A. Goldstein and J. O'Donnell (Eds) 57–86. N.I.D.A., Washington.

Maddux, J.F. and Desmond, D.P. (1982). Residence relocation inhibits opioid dependence. *Archives of General Psychiatry*, **39**, 1313–7.

Marks, I. (1990). Behavioural (non-chemical) addictions. *British Journal of Addiction*, **85**, 1389–94.

Marks, J. (1987). State-rationed drugs. *Druglink*, **2(4)**, 14.

Marlatt, G.A. and George, W.H. (1984). Relapse prevention: introduction and overview of the model. *British Journal of Addiction*, **79**, 261–73.

McLellan, A.T., Luborsky, L., O'Brien, W.G.E. and Druley, K.A. (1983). Predicting response to alcohol and drug abuse treatments. *Archives of General Psychiatry*, **40**, 620–5.

McLellan, A.T., Luborsky, L., Woody, G.E., O'Brien, C.P. and Druley, K.A. (1982). Is treatment for substance abuse effective? *Journal of the American Medical Association*, **247**, 1423–8.

McLellan, A.T., Woody, G.E., Luborsky, L. and Geohl, L. (1988). Is the counsellor an 'active ingredient' in substance abuse rehabilitation? (An examination of treatment success among four counsellors). *Journal of Nervous and Mental Disease*, **176**, 423–30.

Miller, W.R. (1983). Motivational interviewing with problem drinkers. *Behavioural Psychotherapy*, **11**, 147–72.

Prochaska, J.O. and DiClemente, C.C. (1983). Stages and processes of self-change of smoking: toward an integrative model of change. *Journal of Consulting and Clinical Psychology*, **51**, 390–5.

Rathod, N. (1987). Substitution is not a solution. *Druglink*, **29(4)**, 16.

Riley, D. (1987). The management of the pregnant drug addict. *Bulletin of the Royal College of Psychiatrists*, **11**, 362–5.

Robertson, J.R., Bucknall, A.B.V., Welsby, P.D., Roberts, J.J.K., Inglis, J.M., Peutherer, J.F. and Brettle, R.P. (1986). Epidemic of AIDS-related virus infection among intravenous drug abusers. *British Medical Journal*, **292**, 527–9.

Robson, P. (1992). Opiate misusers; are treatments effective? in *Practical Problems in Clinical Psychiatry*, Hawton, K. and Cowen, P. (Eds). Oxford University Press.

Rounsaville, B.J., Glazer, W., Wiber, C.H., Weissman, M.M. and Kleber, H.D. (1983). Short-term interpersonal psychotherapy in methadone-maintained opiate addicts. *Archives of General Psychiatry*, **40**, 629–36.

Simpson, D. (1979). The relation of time spent in drug abuse treatment to post-treatment outcome. *American Journal Of Psychiatry*, **136**, 1449–53.

Skidmore, C.A., Robertson, J.R. and Roberts, J.J.K. (1989). Changes in HIV risk taking behaviour in intravenous drug users: a second follow-up. *British Journal of Addiction*, **84**, 695–6.

Stewart, J., de Wit, H. and Eikelboom, R. (1984). Role of unconditioned and conditioned drug effects in the self-administration of opiates and stimulants. *Psychological Review*, **91**, 251–68.

Stimson, G.V., Alldritt, L.J., Dolan, K.A., Donoghue, M.C. and Lart, R.A. (1988). Injecting equipment exchange schemes. *Final Report of Monitoring Research Group*, Sociology Department, Goldsmiths' College, London SE14 6NW.

Swift, W., Williams, G., Neill, O. and Grenyer, B. (1990). The prevalence of minor psychopathology in opioid users seeking treatment. *British Journal of Addiction*, **85**, 629–34.

Wolk, J., Wodak, A., Guinan, J.J., Macaskill, P., Simpson, J.M. (1990). The effect of a needle and syringe exchange on a methadone maintenance unit. *British Journal of Addiction*, **85**, 1445–50.

Woody, G.E., Luborsky, L., McLellan, A.T., O'Brien, C.P., Beek, A.T., Blaine, J., Herman, I. and Hole, A. (1983). Psychotherapy for opiate addicts—does it help? *Archives of General Psychiatry*, **40**, 639–45.

Chapter 12; Drug policy—a need for change?

Advisory Council of Misuse of Drugs. (1988). *AIDS and Drug Misuse*. D.H.S.S., H.M.S.O., London.

Alexander, B.K. (1990). Alternatives to the war on drugs. *Journal of Drug Issues*, **20**, 1–27.

Anon (Editorial). (1987). Management of drug addicts: hostility, humanity, and pragmatism. *Lancet*, **1**, 1068–1069.

Berridge, V. (1984). Drugs and social policy; the establishment of drug control in Britain 1900–1930. *British Journal of Addiction*, **79**, 17–29.

Berridge, V. and Rawson, N.S.B. (1979). Opiate use and legislative control; a nineteenth century case study. *Social Science and Medicine*, **13**, 351–63.

Clark, A. (1993). Adding up the pros and cons of legalisation. *International Journal of Drug Policy*, **4**, 116–21.

Engelsman, E.L. (1989). Dutch policy on the management of drug-related problems. *British Journal of Addiction*, **84**, 211–8.

Engelsman, E.L. (1991). Drug misuse and the Dutch. *British Medical Journal*, **302**, 484–5.

Farrell, M., Strang, J. (1990). The lure of masterstrokes: drug legalisation. *British Journal of Addiction*, **85**, 5–7.

Gerada, C., Orgel, M. and Strang, J. (1992). Health clinics for problem drug misusers. *Health Trends*, **24**, 68–69.

Glanz, A. (1986). Findings of a national survey of the role of general practitioners in the treatment of opiate misuse: views on treatment. *British Medical Journal*, **293**, 543–5.

Home Office, (1986). *Tackling drug misuse: a summary of the government's strategy*, (2nd Edition), Home Office, London SW1H 9AT.

Marks, J. (1985). Opium, the religion of the people. *Lancet*, **2**, 1439–1440.

Nadelman, E. (1992). Legalisation or harm reduction: the debate continues. *International Journal of Drug Policy*, **3**, 76–82.

Robson, P. (1992). Illegal drugs; time to change the law? *Oxford Medical School Gazette*, **43**, 20–23.

Ruttenber, A.J. (1991). Stalking the elusive designer drugs:techinques for monitoring new problems in drug abuse. *Journal of Addictive Diseases*, **11**, 71–77.

Saunders, J.B. (1990). The great legalisation debate. *Drug & Alcohol Review*, **9**, 3–5.
Smee, C., Parsonage, M., Anderson, R. and Duckworth, S. (1992). The effect of tobacco advertising on tobacco consumption. *Health Trends*, **24**, 111–6.
Stevenson, R. (1991). The case for legalising drugs. *Economic Affairs*, **11**, 14–17.
Smart, C. (1984). Social policy and drug addiction: a critical study of policy development. *British Journal of Addiction*, **79**, 31–39.
Strang, J. (1987). The prescribing debate. *Druglink*, **2(4)**, 10–12.
Strang, J., Ghodse, H. and Johns, A. (1987). Responding flexibly but not gullibly to drug addiction. *British Medical Journal*, **295**, 1364.
Willis, J. H. (1987). Reflections. *British Journal of Addiction*, **82**, 1181–2.
Wijngaart, G. J. Van de. (1989). What lessons from the Dutch experience can be applied? *British Journal of Addiction*, **84**, 990–2.

Newspaper reports

Drug policy

The Independent (1992). Editorial: *Soft drugs harsh laws*, 20 January 1992.
The Independent (1993). Mark Handscombe: *Cannabis: why doctors want it to be legal*, 23 February 1993.
The Independent on Sunday (1993). William Leith: *In the war on drugs, why losing wins you votes*, 16 August 1993.
The Observer (1992). Will Self: *Drug dealer by appointment to HM Government*, 13 September 1992.
The Times (1992). Will Self: *Shrinking from psychiatry*, June 15, 1992.
The Times (1992). Richard Ford: *Drug agency calls for cannabis cafes*, 17 July 1992.
The Times (1992). Jonathon Green: *A quarter century on, the high and mighty rally again to the cause*, 17 July 1992.
The Times (1992). *Editorial: Overdue for repeal*, 24 July 1992.

Drugs and crime

Daily Mail (1992). Anon: *Bart Simpson 'drug cards' kill teenagers*, 11 November 1992.
The Independent (1993). Terry Kirby: *Illegal drugs worth more than £500m seized in 1992*, 22 February 1993.

The Independent (1993). Terry Kirby and Timothy Ross: *Customs fear drugs epidemic*, 31 March 1993.
The Independent (1993). Terry Kirby: *Drugs 'at the heart of the crime wave among the young'*, 13 May 1993.
The Independent (1994). Terry Kirby: *Rising drug problem linked to violence*, 20 January 1994.

Index

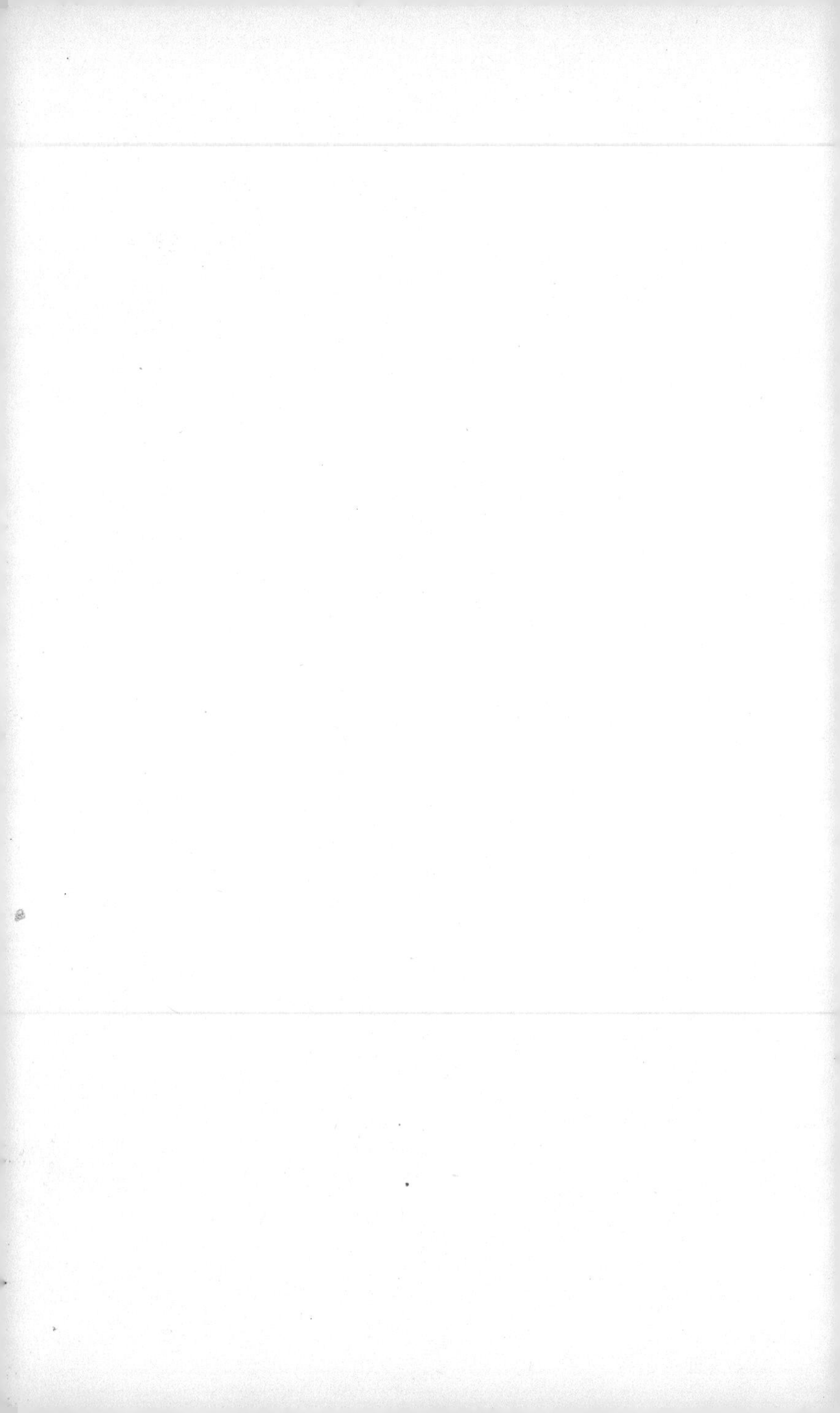

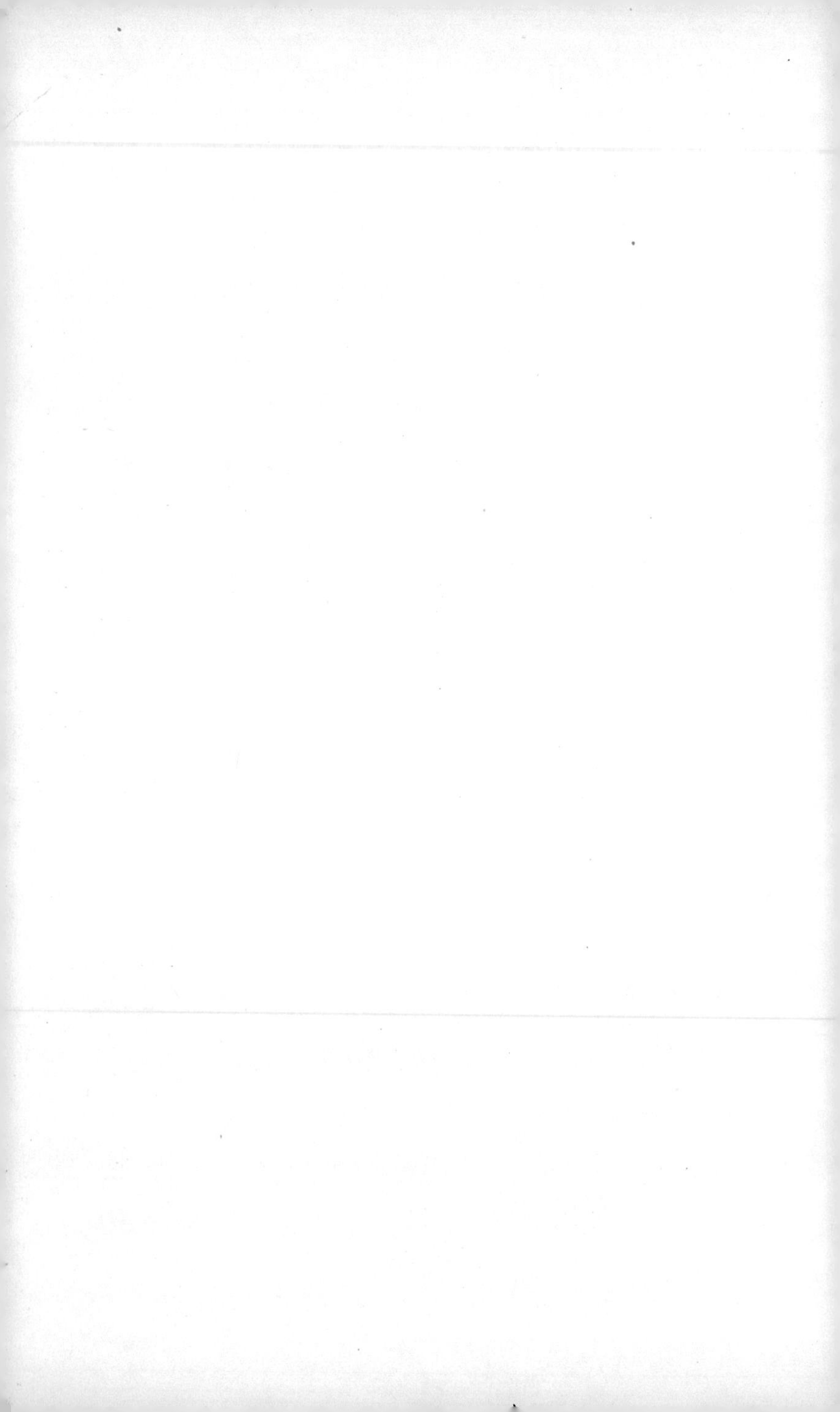